普通高等学校"十四五"规划计算机类专业特色教材

Java Web 项目开发实战

罗　旋　李龙腾　主　编

华中科技大学出版社
中国·武汉

内 容 提 要

全书共分 8 章,以网上书店系统、医院门诊挂号系统、药品库存管理系统、超市商品采购管理系统、物流快递管理系统、旅馆住宿管理系统、火车订票系统和员工管理系统等 8 个实际项目开发程序为案例,从软件工程的角度出发,按照项目的开发顺序,系统、全面地介绍项目开发流程,从系统背景、系统功能需求分析、系统总体功能结构、总体采用的架构、数据库建模到各个功能模块的详细设计与编码实现。

本书适合具有 Java Web 技术基本知识的读者阅读,通过项目案例式引导,以实战带动讲解,让读者能够理论联系实际,提高 Java Web 项目开发实战的能力。

本书的项目开发案例过程完整,可以作为高等院校计算机科学与技术专业、软件工程专业、计算机应用专业,以及其他相关专业的课程设计教材,同时可供从事相关专业的科研人员、软件开发人员及相关大专院校的师生参考。

图书在版编目(CIP)数据

Java Web 项目开发实战/罗旋,李龙腾主编. —武汉:华中科技大学出版社,2021.11
ISBN 978-7-5680-7575-6

Ⅰ.①J… Ⅱ.①罗… ②李… Ⅲ.①JAVA 语言-程序设计 Ⅳ.①TP312.8

中国版本图书馆 CIP 数据核字(2021)第 231911 号

Java Web 项目开发实战　　　　　　　　　　　　　　　　　　　罗　旋　李龙腾　主编
Java Web Xiangmu Kaifa Shizhan

策划编辑:范　莹
责任编辑:陈元玉
封面设计:原色设计
责任监印:徐　露

出版发行:华中科技大学出版社(中国·武汉)　　电话:(027)81321913
　　　　　武汉市东湖新技术开发区华工科技园　　邮编:430223
录　　排:武汉市洪山区佳年华文印部
印　　刷:武汉开心印印刷有限公司
开　　本:787mm×1092mm　1/16
印　　张:16
字　　数:410 千字
版　　次:2021 年 11 月第 1 版第 1 次印刷
定　　价:42.00 元

本书若有印装质量问题,请向出版社营销中心调换
全国免费服务热线:400-6679-118　竭诚为您服务
版权所有　侵权必究

前　　言

在掌握了 Java Web 技术基本知识的基础上，可以通过项目开发来巩固和提高 Java Web 编程技术，本书指导用 Java Web 技术进行实战，开发完整的应用项目。本书将理论融入实践，原理融入技术，突出通用性和实用性，兼具前沿性；以系统化、工程化项目案例的撰写方式，让读者对使用 Java Web 技术开发项目的全过程有一个全面的了解。本书是一本项目开发案例型的、面向应用的软件开发类图书。本书的目的是多角度、全方位地帮助读者快速掌握软件开发技能，架起让学生从学校走向社会的桥梁。

本书选取了网上书店系统、医院门诊挂号系统、药品库存管理系统、超市商品采购管理系统、物流快递管理系统、旅馆住宿管理系统、火车订票系统和员工管理系统等 8 个项目案例。其中，网上书店系统、医院门诊挂号系统、药品库存管理系统和超市商品采购管理系统均基于 MVC 模式设计与开发，运用了 JSP 技术、JavaBean 技术、Servlet 技术和 JDBC 技术；物流快递管理系统、旅馆住宿管理系统和火车订票系统则基于 SSM 框架技术开发，运用了 Spring、Spring MVC、MyBatis 三大框架技术；员工管理系统基于 SpringBoot 框架技术开发。

每章都是以项目为案例，从软件工程的角度出发，按照项目的开发顺序，系统、全面地介绍项目开发流程，从系统背景、系统功能需求分析、系统总体功能结构、总体采用的架构、数据库建模到各个功能模块的详细设计与编码实现，将 Java Web 的主要技术及知识点融入项目案例中，旨在使读者真正获得项目开发经验，提高项目开发能力。

本书各个项目相互独立，读者可以从任何一个项目开始阅读本书，可以按照本书给出的项目开发流程来开发一个软件，也可以参考与这些项目设计类似的软件。

本书以项目贯穿全书，采用模块分解的方式，营造真实的软件企业开发情境，适用于项目教学或理论、实践一体化教学，强化技能训练，提高实战能力，让读者在反复动手的实践过程中学会如何应用所学知识解决实际问题。

本书以完成中小型项目为目的，让学生切身感受到软件开发给他们带来的实实在在的用处和方便，激发学生开发软件的兴趣，调动学生学习的积极性，引导他们根据实际需求，训练自己实际分析问题的能力及编程能力，并养成良好的编程习惯。

由于篇幅有限，本书没有逐一介绍案例中的各模块。笔者选择了基础和典型的模块进行介绍，对于功能重复的模块，由于技术、设计思路和实现过程基本相同，因此没有在书中体现出来。本书提供了所有项目的源代码供读者学习参考，所有程序均经过了笔者精心的调试。需要代码的读者，可以与出版社联系。

本书由罗旋、李龙腾主编。其中，罗旋编写第 5~8 章，李龙腾编写第 1~4 章。全书由罗旋统稿。

本书的项目开发案例过程完整,可以作为高等院校计算机科学与技术专业、软件工程专业、计算机应用专业,以及其他相关专业的课程设计教材,同时可供从事相关专业的科研人员、软件开发人员及相关大专院校的师生参考。

本书引用了许多专家、学者、技术同行的研究成果,在此特向他们表示衷心的感谢。

由于时间仓促和水平有限,书中的疏漏和不妥之处在所难免,敬请读者批评指正。

编 者
2021 年 5 月于武汉

目 录

第1章 网上书店系统 (1)
1.1 需求分析 (1)
1.1.1 系统概述 (1)
1.1.2 功能需求描述 (1)
1.2 总体设计 (2)
1.2.1 系统总体功能结构 (2)
1.2.2 总体架构 (2)
1.2.3 数据库设计 (5)
1.3 详细设计 (6)
1.3.1 用户注册 (6)
1.3.2 信息修改 (7)
1.3.3 图书购买 (7)
1.3.4 图书评论 (8)
1.3.5 添加图书 (9)
1.3.6 图书管理 (9)
1.3.7 用户管理 (11)
1.3.8 售卖记录 (11)
1.3.9 图书列表 (12)
1.4 编码实现 (13)
1.4.1 公共模块 (13)
1.4.2 用户注册 (17)
1.4.3 图书评论 (25)
1.4.4 图书管理 (28)
1.4.5 售卖记录 (32)
1.5 项目搭建 (34)

第2章 医院门诊挂号系统 (36)
2.1 需求分析 (36)
2.1.1 系统概述 (36)
2.1.2 功能需求描述 (36)
2.2 总体设计 (37)
2.2.1 系统总体功能结构 (37)
2.2.2 总体架构 (37)
2.2.3 数据库设计 (38)

2.3 详细设计 (39)
    2.3.1 登录功能 (39)
    2.3.2 挂号预约功能 (40)
    2.3.3 查询、更改、删除功能 (40)
2.4 编码实现 (43)
    2.4.1 患者挂号预约功能 (43)
    2.4.2 医生查询功能 (44)
    2.4.3 医生更改患者信息功能 (45)
2.5 项目搭建 (46)

第3章 药品库存管理系统 (47)

3.1 需求分析 (47)
    3.1.1 系统概述 (47)
    3.1.2 功能需求描述 (47)
3.2 总体设计 (47)
    3.2.1 系统总体功能结构 (47)
    3.2.2 总体架构 (48)
    3.2.3 数据库设计 (49)
3.3 详细设计 (51)
    3.3.1 药品入库 (51)
    3.3.2 药品出库 (52)
    3.3.3 职工信息管理 (53)
3.4 编码实现 (54)
    3.4.1 药品入库管理 (54)
    3.4.2 药品出库管理 (57)
3.5 项目搭建 (61)

第4章 超市商品采购管理系统 (62)

4.1 需求分析 (62)
    4.1.1 系统概述 (62)
    4.1.2 功能需求描述 (62)
4.2 总体设计 (63)
    4.2.1 总体功能结构 (63)
    4.2.2 总体架构 (63)
    4.2.3 数据库设计 (65)
4.3 详细设计 (67)
    4.3.1 采购员 (67)
    4.3.2 超市库存管理员 (69)
    4.3.3 营业员 (70)
    4.3.4 管理员（经理） (71)

4.4 编码实现 …………………………………………………………………… (73)
  4.4.1 采购管理 ……………………………………………………………… (73)
  4.4.2 供货商管理 …………………………………………………………… (78)
  4.4.3 购物缴费 ……………………………………………………………… (83)
  4.4.4 产品列表 ……………………………………………………………… (91)
4.5 项目搭建 …………………………………………………………………… (97)

第5章 物流快递管理系统 ………………………………………………… (98)
5.1 需求分析 …………………………………………………………………… (98)
  5.1.1 系统概述 ……………………………………………………………… (98)
  5.1.2 功能需求描述 ………………………………………………………… (99)
5.2 总体设计 …………………………………………………………………… (100)
  5.2.1 系统总体功能结构 …………………………………………………… (100)
  5.2.2 总体架构 ……………………………………………………………… (102)
  5.2.3 数据库设计 …………………………………………………………… (103)
5.3 详细设计 …………………………………………………………………… (106)
  5.3.1 订单管理 ……………………………………………………………… (106)
  5.3.2 在线下单 ……………………………………………………………… (107)
  5.3.3 查询订单 ……………………………………………………………… (108)
5.4 编码实现 …………………………………………………………………… (109)
  5.4.1 公共模块 ……………………………………………………………… (109)
  5.4.2 订单管理 ……………………………………………………………… (118)
  5.4.3 在线下单 ……………………………………………………………… (129)
  5.4.4 查询订单 ……………………………………………………………… (135)
5.5 项目搭建 …………………………………………………………………… (137)

第6章 旅馆住宿管理系统 ………………………………………………… (139)
6.1 需求分析 …………………………………………………………………… (139)
  6.1.1 系统概述 ……………………………………………………………… (139)
  6.1.2 功能需求描述 ………………………………………………………… (139)
6.2 总体设计 …………………………………………………………………… (140)
  6.2.1 系统总体功能结构 …………………………………………………… (140)
  6.2.2 总体架构 ……………………………………………………………… (141)
  6.2.3 数据库设计 …………………………………………………………… (141)
6.3 详细设计 …………………………………………………………………… (144)
  6.3.1 房间管理 ……………………………………………………………… (144)
  6.3.2 房型管理 ……………………………………………………………… (144)
  6.3.3 入住管理 ……………………………………………………………… (146)
6.4 编码实现 …………………………………………………………………… (147)
  6.4.1 房型管理 ……………………………………………………………… (147)

6.4.2　房间管理 …………………………………………………………………… (158)
　　6.4.3　入住管理 …………………………………………………………………… (167)
6.5　项目搭建 …………………………………………………………………………… (181)

第7章　火车订票系统 ……………………………………………………………… (183)
7.1　需求分析 …………………………………………………………………………… (183)
　　7.1.1　系统概述 ……………………………………………………………………… (183)
　　7.1.2　功能需求描述 ………………………………………………………………… (183)
7.2　总体设计 …………………………………………………………………………… (183)
　　7.2.1　系统总体功能结构 …………………………………………………………… (183)
　　7.2.2　总体架构 ……………………………………………………………………… (184)
　　7.2.3　数据库设计 …………………………………………………………………… (184)
7.3　详细设计 …………………………………………………………………………… (184)
　　7.3.1　车次管理 ……………………………………………………………………… (184)
　　7.3.2　订单管理 ……………………………………………………………………… (185)
7.4　编码实现 …………………………………………………………………………… (187)
　　7.4.1　车次管理 ……………………………………………………………………… (187)
　　7.4.2　订单管理 ……………………………………………………………………… (200)
7.5　项目搭建 …………………………………………………………………………… (210)

第8章　员工管理系统 ……………………………………………………………… (212)
8.1　需求分析 …………………………………………………………………………… (212)
　　8.1.1　系统概述 ……………………………………………………………………… (212)
　　8.1.2　功能需求描述 ………………………………………………………………… (212)
8.2　总体设计 …………………………………………………………………………… (212)
　　8.2.1　系统总体功能结构 …………………………………………………………… (212)
　　8.2.2　总体架构 ……………………………………………………………………… (212)
　　8.2.3　数据库设计 …………………………………………………………………… (214)
8.3　详细设计 …………………………………………………………………………… (216)
　　8.3.1　用户管理 ……………………………………………………………………… (216)
　　8.3.2　角色管理 ……………………………………………………………………… (218)
　　8.3.3　部门管理 ……………………………………………………………………… (219)
　　8.3.4　员工管理 ……………………………………………………………………… (220)
　　8.3.5　日志管理 ……………………………………………………………………… (223)
8.4　编码实现 …………………………………………………………………………… (224)
　　8.4.1　角色管理 ……………………………………………………………………… (224)
　　8.4.2　部门管理 ……………………………………………………………………… (231)
　　8.4.3　员工管理 ……………………………………………………………………… (236)
8.5　项目搭建 …………………………………………………………………………… (246)

参考文献 ……………………………………………………………………………… (247)

第 1 章　网上书店系统

1.1　需求分析

1.1.1　系统概述

　　本章介绍网上书店系统,系统用户划分为管理员用户和普通用户。管理员用户可以进行用户管理,包括查看、删除、修改注册用户信息,可以进行图书管理,包括查看、查询、添加、修改和删除图书信息,还可进行售卖记录查询。普通用户可进行图书查看、购买、评论等操作。为方便日常运营,管理员必须能够简单而便捷地进行图书的上架、下架,以及图书信息的修改等操作,为了能够明确地了解线上销售情况,需要提供一个已售产品清单。销售清单需要显示每种图书已售出多少,库存还剩多少,便于店主有针对性地进货与推出新的产品。为了直接显示线上书店的销售情况,可以自动计算出共入账了多少元。为了避免异常账号干扰正常经营,管理员必须能够对用户账号进行删除操作。面向用户的页面首先必须有注册选项,便于新用户的加入。登录账户的设置是便于用户能够了解自己的购买历史和在该书店的消费金额,便于店主发现优质客户,科学管理,提高效益,也是为后期线上书店的完善提供数据。为了方便用户能够快速地定位到自己心仪的图书,需要提供关于图书的分类管理,通过划分种类,可以提高用户的体验感,还可以为用户提供同类图书的推荐。由于用户信息的不确定性,所以用户还必须可以随意修改自己的注册信息。为了方便我们与用户交流、用户与用户的交流,以及发现用户对于不同图书的建议,因此,在每本书的详细页面还需提供留言栏,以便查看用户的留言。

1.1.2　功能需求描述

　　用户注册:在系统登录页面单击"注册新用户",可进入用户注册页面,填写用户名、密码、年龄、电话等信息后提交,若输入信息为空或格式不正确,则无法注册。

　　用户修改:用户登录后,单击导航栏"修改个人信息",可以进入用户修改页面,用户可修改密码、年龄、电话号码等信息,若输入信息为空或格式不正确,则无法修改。

　　用户删除:管理员进入用户管理页面,可删除用户账号。

　　图书添加:管理员进入图书添加页面,填写图书信息(名称、单价、种类、库存、描述),提交后可添加图书,若图书信息为空或填写不正确,则无法添加。

　　图书管理:管理员进入图书管理页面,可更改图书的价格、库存等信息。

　　图书列表:用户登录后,可通过导航栏"图书列表"功能进入图书列表页面,用户可查看图书列表及基本信息,可以浏览全部图书,也可按类型浏览图书。

　　图书购买:用户已注册并成功登录后,可单击导航栏"图书购买"进入图书购买页面,用户输入对应图书的购买数量,若图书有库存,且购买数量不超过库存数量,则可完成购买,若图书库存为"0",系统在图书列表显示"已售罄",则隐藏购买按钮。

图书评论:用户登录后,可单击导航栏"图书评论"进入图书评论页面,输入相应的评论,系统自动更新评论列表。

查看售卖记录:管理员进入售卖记录页面,系统调用数据库,显示售卖记录详情。

1.2 总体设计

1.2.1 系统总体功能结构

依据需求分析结果,网上书店系统大致可分为数据库访问模块、用户管理模块、图书管理模块、图书交易模块、图书评论模块等5个模块,如图1-1所示。

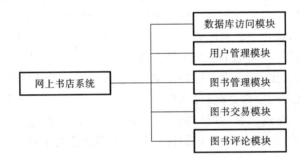

图1-1 网上书店系统

1. 数据库访问模块

数据库访问模块负责处理系统中关于数据库访问的公共操作,主要实现数据库连接、数据库关闭、执行查询类语句、执行非查询类语句等功能。

2. 用户管理模块

用户管理模块负责管理系统中所有合法注册用户的信息,并执行用户的登录、退出等操作。该模块主要实现用户注册、用户登录、修改用户信息、用户删除等功能。

3. 图书管理模块

图书管理模块负责管理系统中的所有图书信息。该模块实现查看、添加、修改、删除图书等功能。

4. 图书交易模块

图书交易模块负责系统中图书的购买下单,以及交易记录查看。该模块实现图书购买和查看售卖记录功能。

5. 图书评论模块

图书评论模块负责用户评论,以及查看评论等。

1.2.2 总体架构

网上书店系统采用MVC架构,包括Model(模型)层、View(视图)层和Controller(控制)层。模型层定义了系统中涉及的各种数据的模型和接口,视图层提供用户交互的页面,控制层提供相关的业务逻辑控制。总体架构如图1-2所示。

MVC架构		
Model层	View层	Controller层
BasicJDBC.java	login.jsp	UserServlet.java
EncondingFilter.java	failure.jsp	BookServlet.java
Book.java	addbook.jsp	MessageServlet.java
Message.java	bookmanage.jsp	SaleServlet.java
Sale.java	listbook.jsp	
Type.java	manage.jsp	
User.java	salebook.jsp	
BookDao.java	salemanage.jsp	
BookDaoImpl.java	user.jsp	
MessageTypeDao.java	usercreate.jsp	
MessageTypeDaoImpl.java	usermanage.jsp	
SaleDao.java		
SaleDaoImpl.java		
UserDao.java		
UserDaoImpl.java		

图1-2 总体架构图

1. 模型层

BasicJDBC.java:为整个应用程序目标数据库提供一个统一的连接对象。

EncondingFilter.java:为字符编码过滤器。

Book.java:用于封装产品图书的信息。

Message.java:用于封装用户留言信息。

Sale.java:用于封装产品售出信息。

Type.java:用于封装产品图书的种类信息。

User.java:用于封装用户信息。

BookDao.java:用于定义图书表单的显示、图书信息的添加、图书信息的修改的接口。

BookDaoImpl.java:用于实现BookDao接口的类。

MessageTypeDao.java:用于定义用户对图书评价的接口。

MessageTypeDaoImpl.java:用于实现MessageTypeDao接口的类。

SaleDao.java:用于显示书店已售出图书的表单,以及客户已购买的图书表单的接口。

SaleDaoImpl.java:用于实现SaleDao接口的类。

UserDao.java:用于验证用户信息操作、用户注册、用户删除、用户信息修改、用户列表显示的接口。

UserDaoImpl.java:用于实现UserDao接口的类。

2. 视图层

login.jsp：提供用户登录的表单，可以输入用户名和密码，提供注册连接。

failure.jsp：对用户的异常操作进行报错。

addbook.jsp：管理员页面添加产品信息的表单页面，可输入名称、单价、种类、库存量、描述。

bookmanage.jsp：管理员修改图书信息的页面，可修改图书的单价及库存。

listbook.jsp：图书浏览页面，且提供分级浏览。

manage.jsp：管理员提供后台管理的页面，提供如下功能连接：添加图书、修改图书信息、管理用户账号、销售历史清单、显示图书列表。

salebook.jsp：显示图书详情页面，用户可键入对图书的评价。

salemanage.jsp：管理员页面图书的销售清单，可自动计算出总收入。

user.jsp：显示用户的详细信息及修改信息、用户的已购历史及总的消费金额。

usercreate.jsp：用户注册页面，可输入用户名、密码、年龄、电话号码。

usermanage.jsp：管理员对用户账号进行管理的页面，可删除选中的用户账号。

3. 控制层

UserServlet.java：进行登录检查，根据 login.jsp 中提交的用户名和密码，调用 Dao 包中的 UserDao.java 和 UserDaoImpl.java 进行数据库验证，若成功，则跳转到用户页面（若用户序列为 1，则为管理员，跳转到管理员页面），否则跳转到 failure.jsp。

用户注册根据 usercreate.jsp 中提交的注册信息，调用 Dao 包中的 UserDao.java 和 UserDaoImpl.java 进行数据库插入，若成功，则跳转到 login.jsp，否则跳转到 failure.jsp。

所有账号列表根据调用 Dao 包中的 UserDao.java 和 UserDaoImpl.java 进行数据库查询，再将获得的信息重定向到 usermanage.jsp。

管理员对所有账号进行管理，再通过 message.jsp 提交过来的信息，调用 Dao 包中的 UserDao.java 和 UserDaoImpl.java 进行数据库中 user 表的更新，若成功，则刷新当前页面，否则跳转到 failure.jsp。

用户信息的列表及信息的修改是根据调用 Dao 包中的 UserDao.java 和 UserDaoImpl.java 进行数据库中 user 表的查询及更新，若成功，则重定向到当前页面，否则跳转到 failure.jsp。

BookServlet.java：添加图书，根据 addbook.jsp 中提交的图书信息，调用 Dao 包中的 BookDao.java 和 BookDaoImpl.java 进行数据库插入，若成功，则跳转到 addbook.jsp，否则跳转到 failure.jsp。

图书列表根据调用 Dao 包中的 BookDao.java 和 BookDaoImpl.java 进行数据库查询，将数据送入 listbook.jsp。

图书的修改根据调用 Dao 包中的 BookDao.java 和 BookDaoImpl.java 进行数据库查询，将数据送入 bookmessage.jsp，而后执行修改方法，再次调用 Dao 包中的 BookDao.java 和 BookDaoImpl.java 进行数据库中 book 表的更新，若成功，则跳转到 bookmessage.jsp，否则跳转到 failure.jsp。

图书的分级浏览根据调用 Dao 包中的 BookDao.java 和 BookDaoImpl.java 进行数据

库查询分组，然后将数据传入 listbook.jsp。

MessageServlet.java：用户对图书进行的评价，根据提交键入的内容，调用 Dao 包中的 MessageTypeDao.java 和 MessageTypeDaoImpl.java 进行数据库插入及查询显示，若成功，则显示输入的评价，否则跳转到 failure.jsp。

SaleServlet.java：交易记录，根据页面表单提交的请求，调用 Dao 包中的 SaleDao.java 和 SaleDaoImpl.java 进行数据库的查询显示，若成功，则显示交易记录的表单，否则跳转到 failure.jsp。

1.2.3 数据库设计

按照系统需求，数据库中包括图书信息表、评论信息表、用户信息表和售卖记录表，表的具体定义如表 1-1 至 1-4 所示。

表 1-1 图书信息表

名 称	字 段 名 称	数 据 类 型	主 键	非 空
图书 ID	Bid	数字（自动编号）	是	是
书名	Bname	文本	否	是
单价	Bprice	数字	否	是
类型	Btype	文本	否	是
库存	Bstock	数字	否	是
售出	Bsale	数字	否	是
描述	Bdesc	文本	否	是

表 1-2 评论信息表

名 称	字 段 名 称	数 据 类 型	主 键	非 空
评论 ID	Mid	数字	是	是
用户 ID	Uid	数字	否	是
图书 ID	Bid	数字	否	是
评论	Message	文本	否	是

表 1-3 用户信息表

名 称	字 段 名 称	数 据 类 型	主 键	非 空
用户 ID	Uid	数字（自动编号）	是	是
姓名	Uname	文本	否	是
密码	Upassword	文本	否	是
年龄	Uage	数字	否	否
电话	Uphone	文本	否	是
描述	Udesc	文本	否	否

表 1-4 售卖记录表

名　　称	字段名称	数据类型	主　　键	非　　空
售卖记录 ID	Sid	数字	是	是
用户 ID	Uid	数字	是	是
图书 ID	Bid	数字	是	是
销售数量	Scount	数字	否	是

1.3 详细设计

1.3.1 用户注册

对网上书店系统的新用户进行记录，给予其访问系统的身份。登录页面如图 1-3 所示，其顺序图如图 1-4 所示。

在数据库中创建的用户信息表 user，用于储存用户的用户名、密码、年龄、电话号码等信息，通过 JDBC 与数据库建立连接。定义 JavaBean 类 User，封装用户信息编写 UserDao 接口并定义 userCreate() 方法，编写 UserDaoImpl 类实现 UserDao 接口。

创建 usercreate.jsp 页面，用户在该页面的文本框中输入注册信息后，将信息提交到 UserServlet 再调用 userCreate()，userCreate() 接收信息并调用 Dao 层（UserDaoImpl 类）的 userCreate() 方法，然后将用户注册信息存入数据库，并返回布尔值来作为注册成功的标

图 1-3 登录页面

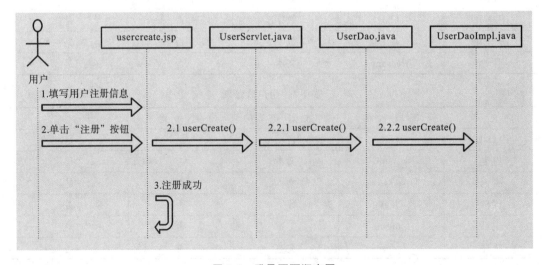

图 1-4 登录页面顺序图

识,若为1,则注册成功返回登录页面,若为0,则显示操作失败页面。

1.3.2 信息修改

信息修改页面是方便用户后期自行更改密码等信息。信息修改页面如图1-5所示,其顺序图如图1-6所示。

图1-5 信息修改页面

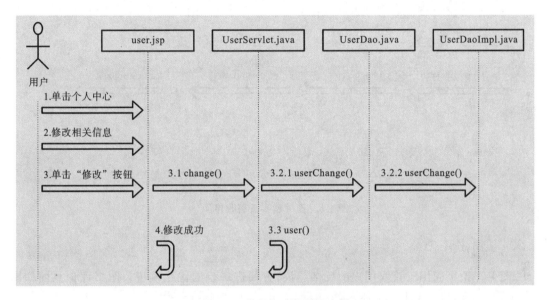

图1-6 信息修改页面顺序图

创建user.jsp页面,用户可在留言框中添加留言信息。将表单以post的方式传递给UserServlet中的change()方法,change()方法接收信息并调用Dao层的userChange()方法,执行数据库的更新语句,更新数据库里的用户信息。完成数据库的操作后,关闭与数据库的连接,返回布尔值来作为完成的标识,若返回值为1,则刷新页面显示更改后的信息,若返回值为0,则跳转到操作失败页面。

1.3.3 图书购买

在图书购买页面,用户可以购买图书。该页面如图1-7所示,图书购买顺序图如图1-8所示。

创建salebook.jsp页面,在该页面中设置售卖记录的按钮,当点击事件发生时,将请求传递给SaleServlet中的booksale()方法。booksale()方法接收请求并调用Dao层的saleBook()方法,进入数据库,将售卖记录表传送给表单变量salelist,操作完成后关闭数据库,跳转到salebook.jsp页面显示售卖记录。

图 1-7 图书购买页面

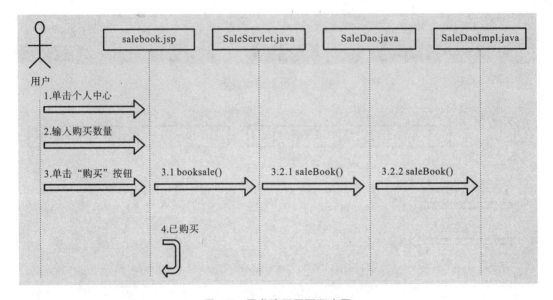

图 1-8 图书购买页面顺序图

1.3.4 图书评论

图书评论页面用于获取用户的反馈建议，方便进一步的优化改进。图书评论页面如图 1-9 所示，其顺序图如图 1-10 所示。

图 1-9 图书评论页面

在数据库中创建用户信息表 message，用于记录留言用户的用户名和留言内容。通过 JDBC 与数据库建立连接。定义 JavaBean 类 Message，封装留言信息，编写 MessageTypeDao 接口并定义 addMessage()方法，编写 MessageTypeDaoImpl 类并实现 MessageTypeDao 接口。

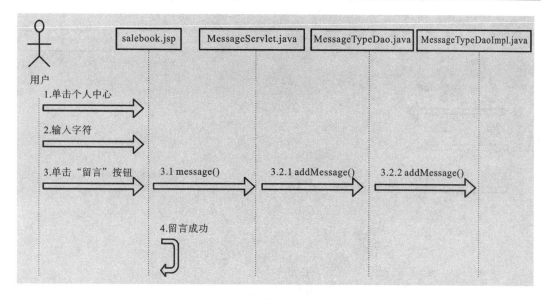

图 1-10 图书评论页面顺序图

创建 usercreate.jsp 页面,用户在该页面的文本框里输入留言信息后,再将信息提交到 Message-Servlet 并调用 message()方法,message()方法接收信息并调用 Dao 层(MessageTypeDaoImpl 类)的 addMessage()方法,然后将用户留言信息存入数据库,并返回布尔值作为注册成功的标志,若返回值为 1,则刷新页面显示留言信息,若返回值为 0,则显示操作失败页面。

1.3.5 添加图书

添加图书页面是指管理员对即将入库的图书的信息进行录入,其页面如图 1-11 所示,其顺序图如图 1-12 所示。

创建 addbook.jsp 页面,在该页面中建立表单,记录入库图书的信息。将该表单以 post 的方式传递给 BookServlet 中的 add()方法,add()方法接收信息并调用 Dao 层的 addBook()方法,再将录入的图书信息存入数据库 book 表中,并返回布尔值来作为注册成功的标志,若返回值为 1,则表示添加成功,更新 addbook.jsp 页面,若返回值为 0,则显示操作失败页面。

1.3.6 图书管理

图书管理页面实施的对象为管理员,管理员可以对图书的库存量和价格进行更改。图书管理页面如图 1-13 所示,其顺序图如图 1-14 所示。

图 1-11 添加图书页面

创建 bookmanage.jsp 页面,在该页面中建立表单,记录所有图书的价格和库存。将该表单以 post 的方式传递给 BookServlet 中的 change()方法,

图 1-12　添加图书页面顺序图

名称	单价	种类	库存	修改
Java Web应用开发教程	28.0	计算机	17	修改
大学英语1	25.8	英语	49	修改
Linux教程	36.0	计算机	29	修改

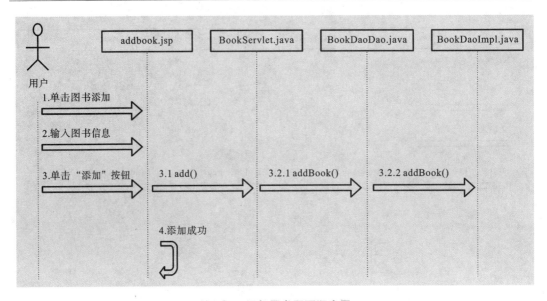

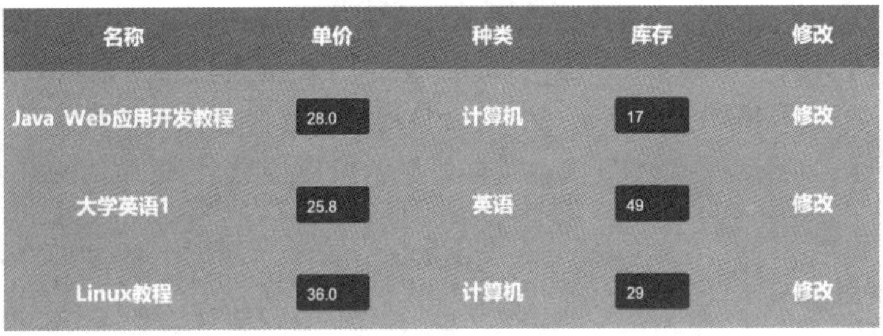

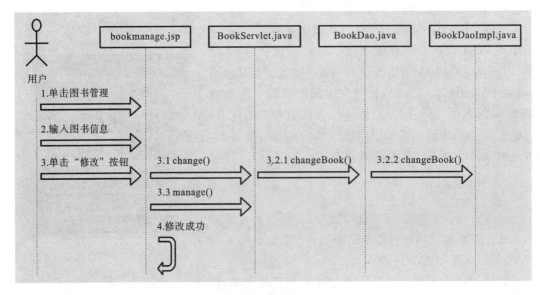

图 1-13　图书管理页面

图 1-14　图书管理页面顺序图

change()方法接收信息并调用 Dao 层的 changeBook()方法,执行数据库的更新语句,更新数据库中图书的价格和库存。完成数据库的操作后,关闭与数据库的连接。返回布尔值来作为完成的标志,若返回值为 1,则刷新页面显示更改后的信息,否则跳转到操作失败页面。

1.3.7 用户管理

用户管理页面是删除已注销用户的信息(管理员操作)。用户管理页面如图 1-15 所示,其顺序图如图 1-16 所示。

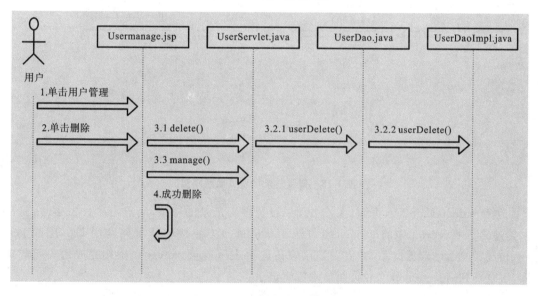

图 1-15 用户管理页面

图 1-16 用户管理页面顺序图

创建 usermanage.jsp 页面,在该页面中创建表单用于存放用户的基本信息。将该表单以 post 的方式传递给 UserServlet 中的 delete()方法,delete()方法接收信息并调用 Dao 层的 userDelete()方法,在数据库 user 表中检索要删除的用户 Uid,找到后将其 Uid 列设为 1(默认值为 0,设为 1 后将无法登录系统并在用户列表中不显示,达到删除目的)。完成数据库的操作后,关闭与数据库的连接。返回布尔值来作为完成的标志,若返回值为 1,则刷新页面,删除用户不再显示在用户列表中,若返回值为 0,则跳转到操作失败页面。

1.3.8 售卖记录

管理员可查看图书的售卖记录,以便购进符合用户观看的图书。售卖记录模块页面如

图 1-17 所示,其顺序图如图 1-18 所示。

用户名	书籍名	数量	单价	总额
admin	Java Web应用开发教程	1	28.0	28.0
admin	Java Web应用开发教程	1	28.0	28.0
admin	Java Web应用开发教程	1	28.0	28.0

图 1-17 售卖记录模块页面

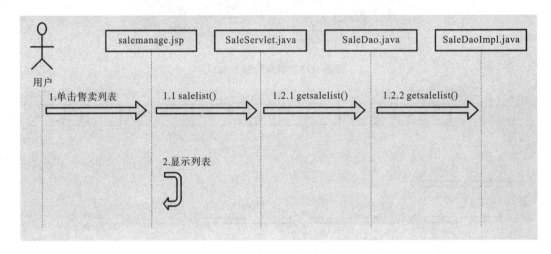

图 1-18 售卖记录模块页面顺序图

创建 salemanage.jsp 页面,在该页面中设置售卖记录的按钮,当点击事件发生时,将请求传递给 SaleServlet 中的 salelist()方法。salelist()方法接收请求并调用 Dao 层的 getsalelist()方法,进入数据库,将售卖记录表传递给表单变量 salelist。操作完成后关闭数据库,再跳转到 salemanage.jsp 页面显示售卖记录。

1.3.9 图书列表

可以给用户展示图书列表及基本信息。用户可以浏览全部图书,也可按类型浏览图书。图书列表页面如图 1-19 所示,其顺序图如图 1-20 所示。

在数据库中创建图书表 book,用于存放图书的基本信息。通过 JDBC 与数据库建立连接。定义 JavaBean 类 Message 和 Type,封装图书的基本信息以及图书类型信息。

创建 listbook.jsp 页面,编辑两个表单。

表单 1 将信息提交到 BookServlet 并调用 list()方法,list()方法接收信息并调用 Dao 层,进入数据库,获取图书信息后传递给表单 booklist 和 typelist,在 listbook.jsp 页面中显示信息。

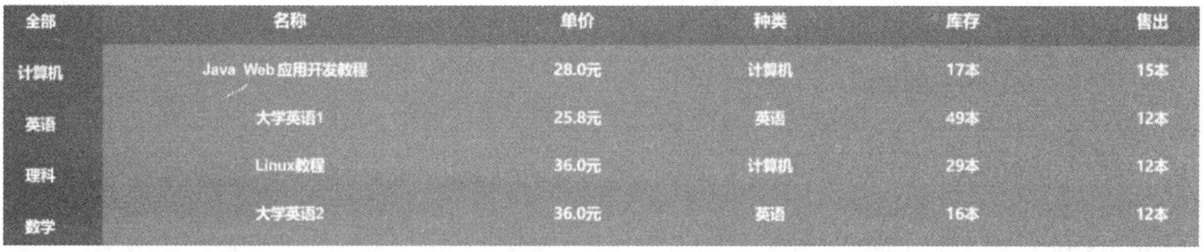

全部	名称	单价	种类	库存	售出
计算机	Java Web应用开发教程	28.0元	计算机	17本	15本
英语	大学英语1	25.8元	英语	49本	12本
理科	Linux教程	36.0元	计算机	29本	12本
数学	大学英语2	36.0元	英语	16本	12本

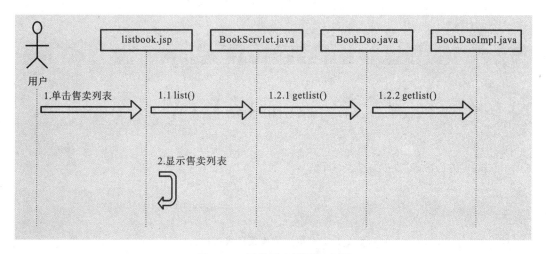

图 1-19　图书列表页面

图 1-20　图书列表页面顺序图

表单 2 将信息提交到 BookServlet 并调用 type()方法，type()方法接收信息并调用 Dao 层，进入数据库，执行查询语句，将图书按照所选类型检索出来，再将其信息传递给表单 booklist 和 typelist。在 listbook. jsp 页面中显示信息。

1.4　编码实现

本节重点介绍公共模块、用户注册、图书评论、图书管理和售卖记录等内容。

1.4.1　公共模块

1. 数据库连接

数据库连接通过 com. util 包下的 BasicJDBC. java 为整个应用程序目标数据库提供统一的连接对象。在连接数据库之前，首先加载要连接数据库的驱动程序到 JVM(Java 虚拟机)。加载到 JDBC 驱动程序通过 java. lang. Class 类的静态方法 forName(String className)实现：通过 DriverManager 类的静态方法 getConnection(String url、String user、String password)建立数据的连接。当构建 Connection、Statement 和 ResultSet 实例时，均需占用一定的数据库和 JDBC 资源，所以每次访问资源结束后，应通过各实例的 close()方法释放它们所占用的资源。

BasicJDBC. java 文件的源代码如下：

```java
package com.util;
import java.sql.Connection;
import java.sql.DriverManager;
import java.sql.PreparedStatement;
import java.sql.ResultSet;
import java.sql.SQLException;
public class BasicJDBC {
    private Connection con=null;
    public BasicJDBC() {
        try {
            Class.forName("com.mysql.cj.jdbc.Driver");
            System.out.println("加载数据库驱动程序成功!");
        } catch (ClassNotFoundException e) {
            e.printStackTrace();
            System.out.println("加载数据库驱动程序失败!");
        }
    }
    public Connection getCon() {
        try {
            con=DriverManager.getConnection("jdbc:mysql://localhost:
                3306/bookshop", "root","1234");
            System.out.println("获取数据库连接成功!");
        } catch (SQLException e) {
            e.printStackTrace();
            System.out.println("获取数据库连接失败!");
        }
        return con;
    }
    public void closeAll(Connection con,PreparedStatement ps,ResultSet rs) {
        if (rs !=null) {
            try {
                rs.close();
            } catch (SQLException e) {
                e.printStackTrace();
            }
        }
        if (ps !=null) {
            try {
                ps.close();
            } catch (SQLException e) {
                e.printStackTrace();
            }
        }
        if (con !=null) {
```

```
        try {
            con.close();
        } catch (SQLException e) {
            e.printStackTrace();
        }
    }
}
```

2. 过滤器

通过 com.lx.filter 包下的 EncodingFilter.java 实现过滤功能。Filter 程序是一个实现了特殊接口的 Java 类,与 Servlet 类似,也是由 Servlet 容器进行调用和执行的。当在 web.xml 中注册一个 Filter 来对某个 Servlet 程序进行拦截处理时,Filter 可以决定是否将请求继续传递给 Servlet 程序,以及对请求和响应消息是否进行修改。当 Servlet 容器开始调用某个 Servlet 程序时,如果发现已经注册一个 Filter 程序来对该 Servlet 程序进行拦截,那么 Servlet 容器不再直接调用 Servlet 的 service() 方法,而是调用 Filter 的 doFilter() 方法,再由 doFilter() 方法决定是否去激活 service() 方法。

EncodingFilter.java 文件的源代码如下:

```
public class EncodingFilter implements Filter {
    public EncodingFilter() {
        System.out.println("过滤器构造");
    }
    public void destroy() {
        System.out.println("过滤器销毁");
    }
    public void doFilter(ServletRequest request,ServletResponse response,
        FilterChain chain) throws IOException,ServletException {
        request.setCharacterEncoding("utf-8");
        response.setContentType("text/html;charset=utf-8");
        chain.doFilter(request,response);
    }
    public void init(FilterConfig arg0) throws ServletException {
        System.out.println("过滤器初始化");
    }
}
```

3. 配置文件

配置文件 web.xml 对不同的对象进行配置,具体代码如下:

```
<?xml version="1.0" encoding="utf-8"?>
<web-app version="2.5"
    xmlns="http://java.sun.com/xml/ns/javaee"
    xmlns:xsi="http://www.w3.org/2001/XMLSchema-instance"
```

```xml
xsi:schemaLocation="http://java.sun.com/xml/ns/javaee
http://java.sun.com/xml/ns/javaee/web-app_2_5.xsd">
<display-name></display-name>
<servlet>
    <description>This is the description of my J2EE component</description>
    <display-name>This is the description of my J2EE component</display-name>
    <servlet-name>BookServlet</servlet-name>
    <servlet-class>com.servlet.BookServlet</servlet-class>
</servlet>
<servlet-mapping>
    <servlet-name>BookServlet</servlet-name>
    <url-pattern>/BookServlet</url-pattern>
</servlet-mapping>
<welcome-file-list>
    <welcome-file>addbook.jsp</welcome-file>
</welcome-file-list>
<servlet>
    <description>This is the description of my J2EE component</description>
    <display-name>This is the description of my J2EE component</display-name>
    <servlet-name>SaleServlet</servlet-name>
    <servlet-class>com.servlet.SaleServlet</servlet-class>
</servlet>
<servlet-mapping>
    <servlet-name>SaleServlet</servlet-name>
    <url-pattern>/SaleServlet</url-pattern>
</servlet-mapping>
<welcome-file-list>
    <welcome-file>salebook.jsp</welcome-file>
</welcome-file-list>
<servlet>
    <description>This is the description of my J2EE component</description>
    <display-name>This is the description of my J2EE component</display-name>
    <servlet-name>MessageServlet</servlet-name>
    <servlet-class>com.servlet.MessageServlet</servlet-class>
</servlet>
<servlet-mapping>
    <servlet-name>MessageServlet</servlet-name>
    <url-pattern>/MessageServlet</url-pattern>
</servlet-mapping>
<welcome-file-list>
    <welcome-file>salebook.jsp</welcome-file>
</welcome-file-list>
<servlet>
    <description>This is the description of my J2EE component</description>
```

```xml
        <display-name>This is the description of my J2EE component</display-name>
        <servlet-name>UserServlet</servlet-name>
        <servlet-class>com.servlet.UserServlet</servlet-class>
    </servlet>
    <servlet-mapping>
        <servlet-name>UserServlet</servlet-name>
        <url-pattern>/UserServlet</url-pattern>
    </servlet-mapping>
    <welcome-file-list>
        <welcome-file>login.jsp</welcome-file>
    </welcome-file-list>
    <filter>
        <filter-name>EncodingFilter</filter-name>
        <filter-class>com.filter.EncodingFilter</filter-class>
    </filter>
    <filter-mapping>
        <filter-name>EncodingFilter</filter-name>
        <url-pattern>/*</url-pattern>
    </filter-mapping>
</web-app>
```

1.4.2 用户注册

通过数据库创建用户信息表 user，用来储存用户的用户名、密码、年龄、电话号码等信息。通过 JDBC 与数据库建立连接。定义 JavaBean 类 User，封装用户信息，编写 UserDao 接口并定义 userCreate() 方法，编写 UserDaoImpl 类来实现 UserDao 接口。

创建 usercreate.jsp 页面，用户在该页面文本框中输入注册信息后，将信息提交到 UserServlet 并调用 userCreate() 方法，userCreate() 方法接收信息并调用 Dao 层（UserDaoImpl 类）的 userCreate() 方法，再将用户注册信息存入数据库。返回布尔值作为注册成功的标志，若返回值为 1，则表示注册成功并返回登录页面，否则显示操作失败页面。

1. Model 层

（1）UserDao.java 文件的源代码如下：

```java
//定义用户类接口 UserDao：
public interface UserDao {
    public int valiLogin(String name,String pwd) throws Exception;
    public int userCreate(String uname,String upassword,int uage,String uphone);
    public int userDelete(int uid);
    public int userChange(int uid,String uname,String upassword,int uage,
        String uphone);
    public List<User> getUser(int uid);
}
```

（2）UserDaoImpl.java 文件的源代码如下：

```java
//定义用户操作类 UserDaoImpl:
public class UserDaoImpl implements UserDao {
    private BasicJDBC db=null;
    private Connection con=null;
    public UserDaoImpl() {
        db=new BasicJDBC();
        con=db.getCon();
    }
    //登录验证
    public int valiLogin(String name,String password) throws Exception {
        int uid=0;
        PreparedStatement ps=null;
        ResultSet rs=null;
        String sql="select * from user where Uname=? and Upassword=? and Ud=0";
        try {
            ps=con.prepareStatement(sql);
            ps.setString(1, name);
            ps.setString(2, password);
            rs=ps.executeQuery();
            if(rs.next()) {
                uid=rs.getInt(1);
            }
        } catch(SQLException e) {
            e.printStackTrace();
        } finally {
            db.closeAll(con, ps, null);
        }
        return uid;
    }
    //创建用户
    public int userCreate(String uname, String upassword, int uage, String uphone) {
        PreparedStatement ps=null;
        String sql="insert into user(Uname,Upassword,Uage,Uphone,Ud) values(?,?,?,?,0)";
        int n=0;
        try {
            ps=con.prepareStatement(sql);
            ps.setString(1, uname);
            ps.setString(2, upassword);
            ps.setInt(3, uage);
            ps.setString(4, uphone);
            n=ps.executeUpdate();
            System.out.println(n);
        } catch (SQLException e) {
            e.printStackTrace();
```

```java
        } finally {
            db.closeAll(con, ps, null);
        }
        return n;
    }
    //根据uid查询用户信息
    public List<User>getUser(int uid) {
        PreparedStatement ps=null;
        String sql="select* from user where Ud=0 and Uid=";
        if (uid==0) {
            sql=sql+"Uid";
        } else {
            sql=sql+uid;
        }
        ResultSet rs=null;
        List<User> userlist=new ArrayList<User>();
        try {
            ps=con.prepareStatement(sql);
            rs=ps.executeQuery();
            while (rs.next()) {
                User u=new User(rs.getInt(1), rs.getString(2),
                    rs.getString(3),rs.getInt(4),rs.getString(5));
                userlist.add(u);
            }
        } catch (SQLException e) {
            e.printStackTrace();
        } finally {
            db.closeAll(con, ps, rs);
        }
        return userlist;
    }
    //删除用户
    public int userDelete(int uid) {
        PreparedStatement ps=null;
        String sql="update user set Ud=1 where Uid=?";
        int n=0;
        try {
            ps=con.prepareStatement(sql);
            ps.setInt(1, uid);
            n=ps.executeUpdate();
            System.out.println(n);
        } catch (SQLException e) {
            e.printStackTrace();
        } finally {
```

```java
            db.closeAll(con, ps, null);
        }
        return n;
    }
    //修改用户
    public int userChange(int uid,String uname,String upassword,int uage,
        String uphone) {
        PreparedStatement ps=null;
        String sql="update user set Uname=?,Upassword=?,Uage=?,
            Uphone=? where Uid=?";
        int n=0;

        try {
            ps=con.prepareStatement(sql);
            ps.setString(1, uname);
            ps.setString(2, upassword);
            ps.setInt(3, uage);
            ps.setString(4, uphone);
            ps.setInt(5, uid);
            n=ps.executeUpdate();
            System.out.println(n);
        } catch (SQLException e) {
            e.printStackTrace();
        } finally {
            db.closeAll(con, ps, null);
        }
        return n;
    }
}
```

2. View 层

用户注册页面源文件 usercreate.jsp 的具体代码如下：

```jsp
<%@ page language="java" contentType="text/html;charset=utf-8"
    pageEncoding="utf-8"%>
<!DOCTYPE html>
<html>
<head>
<meta charset="utf-8">
    <title>用户注册</title>
</head>
<style>
    html{
        width:100%;
        height:100%;
```

```css
    overflow: hidden;
    font-style: sans-serif;
}
body{
    width: 100% ;
    height: 100% ;
    font-family: 'Open Sans',sans-serif;
    margin: 0;
    background-color: #566C73;
}
#login{
    position: absolute;
    top: 50% ;
    left:50% ;
    margin: -225px 0 0 -150px;
    width: 300px;
    height: 300px;
}
#login h1{
    color: #fff;
    font-size: 35px;
    text-shadow:0 0 10px;
    letter-spacing: 1px;
    text-align: center;
}
h1{
    font-size: 2em;
    margin: 0.67em 0;
}
input{
    width: 278px;
    height: 18px;
    margin-bottom: 10px;
    outline: none;
    padding: 10px;
    font-size: 13px;
    color: #fff;
    border-top: 1px solid #312E3D;
    border-left: 1px solid #312E3D;
    border-right: 1px solid #312E3D;
    border-bottom: 1px solid #56536A;
    border-radius: 4px;
    background-color: #2D2D3F;
}
```

```css
.but{
    width: 300px;
    min-height: 20px;
    display: block;
    background-color: #4a77d4;
    border: 1px solid #3762bc;
    color: #fff;
    padding: 9px 14px;
    font-size: 15px;
    line-height: normal;
    border-radius: 5px;
    margin: 0px 0px 10px 0px;
}
</style>
<body>
<div id="login">
    <h1>用户注册</h1>
    <form method="post" action="UserServlet?action=usercreate" method="post">
    <input type="text" required="required" placeholder=
        "用户名" name="uname"></input>
    <input type="password" required="required" placeholder=
        "密码" name="upassword"></input>
    <input type="number" min="0" required="required" placeholder=
        "年龄" name="uage"></input>
    <input type="text" required="required" placeholder=
        "电话" name="uphone"></input>
    <button class="but" type="submit">注册</button>
    <button class="but" type="button" onclick="location.href=
        'login.jsp'">返回登录</button>
    </form>
</div>
</body>
</html>
```

3. Controller 层

UserServlet.java 文件的源代码如下:

```java
public class UserServlet extends HttpServlet {
    private static final long serialVersionUID=1L;
    public UserServlet() {
        super();
    }
    public void init() throws ServletException{}
    public void doGet(HttpServletRequest request,HttpServletResponse response)
            throws ServletException,IOException {
```

```java
        this.doPost(request, response);
}
//处理客户端请求
public void doPost(HttpServletRequest request,HttpServletResponse response)
        throws ServletException,IOException {
    String action=request.getParameter("action");
    if (action.equals("login")) login(request, response);
    if (action.equals("usercreate")) usercreate(request, response);
    if (action.equals("manage")) manage(request, response);
    if (action.equals("delete")) delete(request, response);
    if (action.equals("user")) user(request, response);
    if (action.equals("change")) change(request, response);
}
protected void login(HttpServletRequest request, HttpServletResponse response)
        throws ServletException, IOException {
    String uname=request.getParameter("uname");
    String upassword=request.getParameter("upassword");
    User user=new User(0, uname, upassword, 0, "");
    user.setUname(uname);
    user.setUpassword(upassword);
    String username= (String)user.getUname();
    String userpass= (String)user.getUpassword();
    UserDao userdao=new UserDaoImpl();
    int uid=0;
    try {
        uid=userdao.valiLogin(username, userpass);
        HttpSession session=request.getSession();
        session.setAttribute("uid", uid);
    } catch(Exception e) {
        e.printStackTrace();
        request.getRequestDispatcher("/failure.jsp").forward(request,response);
        return;
    }
    if(uid==0) {
        request.getRequestDispatcher("/failure.jsp").forward(request,response);
    } else if (uid==1) {
        request.getRequestDispatcher("/manage.jsp").forward(request,response);
    } else {
        request.getRequestDispatcher("/BookServlet?action=
            list").forward(request,response);
    }
}
protected void usercreate(HttpServletRequest request, HttpServletResponse response)
        throws ServletException, IOException {
```

```java
        String uname=request.getParameter("uname");
        String upassword=request.getParameter("upassword");
        int uage=Integer.parseInt(request.getParameter("uage"));
        String uphone=request.getParameter("uphone");
        UserDao userdao=new UserDaoImpl();
        int n=0;
        try {
            n=userdao.userCreate(uname, upassword, uage, uphone);
        } catch (Exception e) {
            e.printStackTrace();
            request.getRequestDispatcher("/failure.jsp").forward(request, response);
            return;
        }
        if (n==1) {
            request.getRequestDispatcher("/login.jsp").forward(request, response);
        } else {
            request.getRequestDispatcher("/failure.jsp").forward(request, response);
        }
    }
    protected void manage(HttpServletRequest request, HttpServletResponse response)
            throws ServletException, IOException {
        UserDaoImpl userdao=new UserDaoImpl();
        List<User> userlist=userdao.getUser(0);
        HttpSession session=request.getSession();
        session.setAttribute("userlist", userlist);
        response.sendRedirect("usermanage.jsp");
    }
    protected void delete(HttpServletRequest request, HttpServletResponse response)
            throws ServletException, IOException {
        int uid=Integer.parseInt(request.getParameter("uid"));
        UserDao userdao=new UserDaoImpl();
        int n=0;
        try {
            n=userdao.userDelete(uid);
        } catch (Exception e) {
            e.printStackTrace();
            request.getRequestDispatcher("/failure.jsp").forward(request, response);
            return;
        }
        if (n==1) {
            request.getRequestDispatcher("/UserServlet?action=
                manage").forward(request, response);
        } else {
            request.getRequestDispatcher("/failure.jsp").forward(request, response);
```

```java
            }
        }
        protected void user(HttpServletRequest request, HttpServletResponse response)
                throws ServletException, IOException {
            int uid=Integer.parseInt(request.getParameter("uid"));
            UserDaoImpl userdao=new UserDaoImpl();
            SaleDaoImpl saledao=new SaleDaoImpl();
            List<User> userinfor=userdao.getUser(uid);
            List<Sale> usersale=saledao.getSalelist(uid);
            HttpSession session=request.getSession();
            session.setAttribute("userinfor", userinfor);
            session.setAttribute("usersale", usersale);
            response.sendRedirect("user.jsp");
        }
        protected void change(HttpServletRequest request, HttpServletResponse response)
                throws ServletException, IOException {
            int uid=Integer.parseInt(request.getParameter("uid"));
            String uname=request.getParameter("uname");
            String upassword=request.getParameter("upassword");
            int uage=Integer.parseInt(request.getParameter("uage"));
            String uphone=request.getParameter("uphone");
            UserDao userdao=new UserDaoImpl();
            int n=0;
            try {
                n=userdao.userChange(uid, uname, upassword, uage, uphone);
            } catch (Exception e) {
                e.printStackTrace();
                request.getRequestDispatcher("/failure.jsp").forward(request, response);
                return;
            }
            if (n==1) {
                request.getRequestDispatcher("/UserServlet? action=user").forward(
                    request, response);
            } else {
                request.getRequestDispatcher("/failure.jsp").forward(request, response);
            }
        }
        public void destroy() {
            super.destroy();
        }
    }
```

1.4.3 图书评论

创建 usercreate.jsp 页面，用户在该页面文本框中输入留言信息后，将信息提交到

MessageServlet 并调用 message()方法，message()方法接收信息并调用 Dao 层（MessageTypeDaoImpl 类）的 addMessage()方法，再将用户留言信息存入数据库。返回布尔值作为注册成功的标志，若返回值为 1，则刷新页面显示留言信息，否则显示操作失败页面。

1. Model 层

（1）图书评论接口文件 MessageTypeDao.java 的源代码如下：

```java
//定义图书评论接口 MessageTypeDao
public interface MessageTypeDao {
    public List<Message> getMessageList(int Bid);
    public List<Type> getTypeList();
    public int addMessage(int Uid, int Bid, String Mmessage);
}
```

（2）实现 MessageTypeDao 接口文件 MessageTypeDaoImpl.java 的源代码如下：

```java
//定义图书评论操作类 MessageTypeDaoImpl
public class MessageTypeDaoImpl implements MessageTypeDao {
    private BasicJDBC db=null;
    private Connection con=null;
    public MessageTypeDaoImpl() {
        db=new BasicJDBC();
        con=db.getCon();
    }
    //获取评论列表
    @Override
    public List<Message> getMessageList(int Bid) {
        PreparedStatement ps=null;
        String sql="select Uname,Mmessage from message join user on user.Uid=
            message.Uid where Bid=?";
        ResultSet rs=null;
        List<Message> list=new ArrayList<Message>();
        try {
            ps=con.prepareStatement(sql);
            ps.setInt(1, Bid);
            rs=ps.executeQuery();
            while (rs.next()) {
                Message m=new Message(rs.getString(1), rs.getString(2));
                list.add(m);
            }
        } catch (SQLException e) {
            e.printStackTrace();
        } finally {
            db.closeAll(con, ps, rs);
        }
        return list;
```

```java
}
//添加评论
@Override
public int addMessage(int Uid, int Bid, String Mmessage) {
    PreparedStatement ps=null;
    String sql="insert into message(Uid,Bid,Mmessage) values (?,?,?)";
    int n=0;
    try {
        ps=con.prepareStatement(sql);
        ps.setInt(1, Uid);
        ps.setInt(2, Bid);
        ps.setString(3, Mmessage);
        n=ps.executeUpdate();
        System.out.println(n);
    } catch (SQLException e) {
        e.printStackTrace();
    } finally {
        db.closeAll(con, ps, null);
    }
    return n;
}
```

2. View 层(略)

3. Controller 层

评论处理文件 MessageServlet.java 的源代码如下：

```java
//评论处理 servlet
public class MessageServlet extends HttpServlet {
    private static final long serialVersionUID=1L;
    public MessageServlet() {
        super();
    }
    public void doGet(HttpServletRequest request, HttpServletResponse response)
            throws ServletException, IOException {
        this.doPost(request, response);
    }
    public void doPost(HttpServletRequest request, HttpServletResponse response)
            throws ServletException, IOException {
        message(request, response);
    }
    protected void message(HttpServletRequest request, HttpServletResponse response)
            throws ServletException, IOException {
        int uid=Integer.parseInt(request.getParameter("uid"));
        int bid=Integer.parseInt(request.getParameter("bid"));
```

```
        String mmessage=request.getParameter("mmessage");
        MessageTypeDao messagedao=new MessageTypeDaoImpl();
        int n=0;
        try {
            n=messagedao.addMessage(uid, bid, mmessage);
        } catch (Exception e) {
            e.printStackTrace();
            request.getRequestDispatcher("/failure.jsp").forward(request, response);
            return;
        }
        if (n==1) {
            String url="/SaleServlet?action=booklist";
            request.getRequestDispatcher(url).forward(request, response);
        } else {
            request.getRequestDispatcher("/failure.jsp").forward(request, response);
        }
    }
}
```

1.4.4 图书管理

创建 bookmanage.jsp 页面，在该页面中建立表单，用于记录所有图书的价格和库存。将该表单以 post 的方式传递给 BookServlet 中的 change()方法，change()方法接收信息并调用 Dao 层的 changeBook()方法，执行数据库的更新语句，更新数据库里图书的价格和库存。完成数据库的操作后，关闭与数据库的连接。返回布尔值作为完成的标志，若返回值为1，则刷新页面显示更改后的信息，否则跳转到操作失败页面。

1. Model 层

（1）定义图书类接口文件 BookDao.java 的源代码如下：

```
//定义图书类接口
public interface BookDao {
    public List<Book> getList(String btype);
    public int addBook(Book b);
    public int changeBook(int bid, float bprice, int bstock);
}
```

（2）定义图书操作文件 BookDaoImpl.java 的源代码如下：

```
//定义图书操作类 BookDaoImpl
public class BookDaoImpl implements BookDao {
    private BasicJDBC db=null;
    private Connection con=null;
    public BookDaoImpl() {
        db=new BasicJDBC();
        con=db.getCon();
```

```java
    }
    //添加图书
    @Override
    public int addBook(Book b) {
        PreparedStatement ps=null;
        String sql="insert into book(Bname,Bprice,Btype,Bstock,Bsale,Bdes)
            values (?,?,?,?,?,?)";
        int n=0;
        try {
            ps=con.prepareStatement(sql);
            ps.setString(1, b.getBname());
            ps.setFloat(2, b.getBprice());
            ps.setString(3, b.getBtype());
            ps.setInt(4, b.getBstock());
            ps.setInt(5, b.getBsale());
            ps.setString(6, b.getBdes());
            n=ps.executeUpdate();
            System.out.println(n);
        } catch (SQLException e) {
            e.printStackTrace();
        } finally {
            db.closeAll(con, ps, null);
        }
        return n;
    }
    //修改图书信息
    @Override
    public int changeBook(int bid, float bprice, int bstock) {
        PreparedStatement ps=null;
        String sql="update book set Bprice=?,Bstock=? where Bid=?";
        int n=0;
        try {
            ps=con.prepareStatement(sql);
            ps.setFloat(1, bprice);
            ps.setInt(2, bstock);
            ps.setInt(3, bid);
            n=ps.executeUpdate();
            System.out.println(n);
        } catch (SQLException e) {
            e.printStackTrace();
        } finally {
            db.closeAll(con, ps, null);
        }
        return n;
```

```java
    }
    //根据图书类型查询图书信息
    @Override
    public List<Book> getList(String btype) {
        PreparedStatement ps=null;
        String sql="select * from book where Btype=";
        if (btype=="Btype") {
            sql=sql+btype;
        } else {
            sql=sql+"'"+btype+"'";
        }
        ResultSet rs=null;
        List<Book> list=new ArrayList<Book>();
        try {
            ps=con.prepareStatement(sql);
            rs=ps.executeQuery();
            while (rs.next()) {
                Book b=new Book(rs.getInt(1), rs.getString(2), rs.getFloat(3),
                    rs.getString(4), rs.getInt(5), rs.getInt(6), rs.getString(7));
                list.add(b);
            }
        } catch (SQLException e) {
            e.printStackTrace();
        } finally {
            db.closeAll(con, ps, rs);
        }
        return list;
    }
}
```

2. View 层（略）

3. Controller 层

图书处理 servlet 源文件 BookServlet.java 的源代码如下：

```java
public class BookServlet extends HttpServlet {
    private static final long serialVersionUID=1L;
    public BookServlet() {
        super();
    }
    protected void doGet(HttpServletRequest request, HttpServletResponse response)
            throws ServletException, IOException {
        doPost(request, response);
    }
    protected void doPost(HttpServletRequest request, HttpServletResponse response)
            throws ServletException, IOException {
```

```java
        String action=request.getParameter("action");
        if (action.equals("add")) add(request, response);
        if (action.equals("list")) list(request, response);
        if (action.equals("manage")) manage(request, response);
        if (action.equals("change")) change(request, response);
        if (action.equals("type")) type(request, response);
    }
    //添加图书
    protected void add(HttpServletRequest request, HttpServletResponse response)
            throws ServletException, IOException {
        String bname=request.getParameter("bname");
        Float bprice=Float.parseFloat(request.getParameter("bprice"));
        String btype=request.getParameter("btype");
        int bstock=Integer.parseInt(request.getParameter("bstock"));
        String bdes=request.getParameter("bdes");
        Book book=new Book(0, bname, bprice, btype, bstock, 0, bdes);
        book.setBname(bname);
        book.setBprice(bprice);
        book.setBtype(btype);
        book.setBstock(bstock);
        book.setBdes(bdes);
        BookDao bookdao=new BookDaoImpl();
        int n=0;
        try {
            n=bookdao.addBook(book);
        } catch (Exception e) {
            e.printStackTrace();
            request.getRequestDispatcher("/failure.jsp").forward(request, response);
            return;
        }
        if (n==1) {
            request.getRequestDispatcher("/addbook.jsp").forward(request, response);
        } else {
            request.getRequestDispatcher("/failure.jsp").forward(request, response);
        }
    }
    //修改图书
    protected void change(HttpServletRequest request, HttpServletResponse response)
            throws ServletException, IOException {
        int bid=Integer.parseInt(request.getParameter("bid"));
        Float bprice=Float.parseFloat(request.getParameter("bprice"));
        int bstock=Integer.parseInt(request.getParameter("bstock"));
        BookDao bookdao=new BookDaoImpl();
        int n=0;
```

```java
        try {
            n=bookdao.changeBook(bid, bprice, bstock);
        } catch (Exception e) {
            e.printStackTrace();
            request.getRequestDispatcher("/failure.jsp").forward(request, response);
            return;
        }
        if (n==1) {
            request.getRequestDispatcher("/BookServlet? action=
                manage").forward(request, response);
        } else {
            request.getRequestDispatcher("/failure.jsp").forward(request, response);
        }
    }
    //查询特定类型图书
    protected void type(HttpServletRequest request, HttpServletResponse response)
            throws ServletException, IOException {
        String booktype=request.getParameter("booktype");
        BookDaoImpl bookdao=new BookDaoImpl();
        MessageTypeDao messagetypedao=new MessageTypeDaoImpl();
        List<Book> booklist=bookdao.getList(booktype);
        List<Type> typelist=messagetypedao.getTypeList();
        HttpSession session=request.getSession();
        session.setAttribute("booklist", booklist);
        session.setAttribute("typelist", typelist);
        response.sendRedirect("listbook.jsp");
    }
}
```

1.4.5 售卖记录

创建 salemanage.jsp 页面，在该页面中设置售卖记录的按钮，当点击事件发生时，将请求传递给 SaleServlet 中的 salelist()方法。salelist()方法接收请求并调用 Dao 层的 getsalelist()方法，进入数据库，将售卖记录表传递给表单变量 salelist。操作完成后就关闭数据库，再跳转到 salemanage.jsp 页面显示售卖记录。

1. Model 层

（1）售卖记录接口文件 SaleDao.java 的源代码如下：

```java
//定义售卖记录类接口 SaleDao
public interface SaleDao {
    public List<Book> getBookSale(int bid);
    public List<Sale> getSalelist(int uid);
    public int saleBook(int uid, int bid, int salecount);
}
```

（2）售卖记录操作文件 SaleDaoImpl.java 的源代码如下：

```java
//定义售卖记录操作类 SaleDaoImpl
public class SaleDaoImpl implements SaleDao {
    private BasicJDBC db=null;
    private Connection con=null;
    public SaleDaoImpl() {
        db=new BasicJDBC();
        con=db.getCon();
    }
    //图书销售
    @Override
    public int saleBook(int uid, int bid, int salecount) {
        PreparedStatement ps=null;
        String sql="call sale(?,?,?)";
        int n=0;
        try {
            ps=con.prepareStatement(sql);
            ps.setInt(1, uid);
            ps.setInt(2, bid);
            ps.setInt(3, salecount);
            n=ps.executeUpdate();
            System.out.println(n);
        } catch (SQLException e) {
            e.printStackTrace();
        } finally {
            db.closeAll(con, ps, null);
        }
        return n;
    }
    //获取用户的所有售卖记录
    @Override
    public List<Sale> getSalelist(int uid) {
        PreparedStatement ps=null;
        String sql = "SELECT Uname,Bname,Scount,Bprice FROM sale "
                + "join user on User.Uid=sale.Uid join book on book.Bid="
                    + "sale.Bid where user.Uid=";
        if (uid==0) {
            sql=sql+"User.Uid";
        } else {
            sql=sql+uid;
        }
        ResultSet rs=null;
        List<Sale> salelist=new ArrayList<Sale>();
        try {
```

```java
            ps=con.prepareStatement(sql);
            rs=ps.executeQuery();
            while (rs.next()) {
                Sale s=new Sale(rs.getString(1), rs.getString(2),
                  rs.getInt(3),rs.getFloat(4));
                salelist.add(s);
            }
        } catch (SQLException e) {
            e.printStackTrace();
        } finally {
            db.closeAll(con, ps, rs);
        }
        return salelist;
    }
}
```

2. View 层(略)

3. Controller 层

售卖记录处理 servlet 文件 SaleServlet.java 的源代码如下：

```java
public class SaleServlet extends HttpServlet {
    public void doPost(HttpServletRequest request, HttpServletResponse response)
            throws ServletException,IOException {
        String action=request.getParameter("action");
        if (action.equals("booksale")) booksale(request, response);
        if (action.equals("booklist")) booklist(request, response);
        if (action.equals("salelist")) salelist(request, response);
    }
    //获取售卖记录
    protected void salelist(HttpServletRequest request, HttpServletResponse response)
            throws ServletException, IOException {
        int uid=0;
        SaleDaoImpl saledao=new SaleDaoImpl();
        List<Sale> salelist=saledao.getSalelist(uid);
        HttpSession session=request.getSession();
        session.setAttribute("salelist", salelist);
        response.sendRedirect("salemanage.jsp");
    }
}
```

1.5 项目搭建

网上书店系统使用 eclipse 工具和 MySQL 数据库开发，系统项目结构如图 1-21、图 1-22

所示。

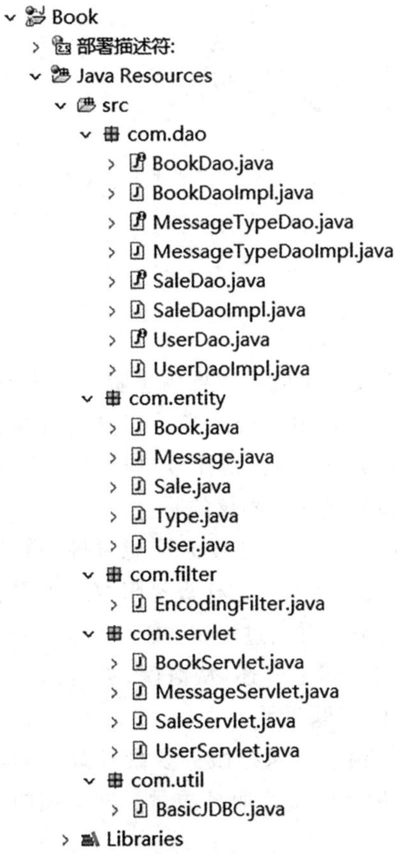

```
v 🗁 Book
  > 📄 部署描述符:
  v 📂 Java Resources
    v 📁 src
      v ⊞ com.dao
        > 📄 BookDao.java
        > 📄 BookDaoImpl.java
        > 📄 MessageTypeDao.java
        > 📄 MessageTypeDaoImpl.java
        > 📄 SaleDao.java
        > 📄 SaleDaoImpl.java
        > 📄 UserDao.java
        > 📄 UserDaoImpl.java
      v ⊞ com.entity
        > 📄 Book.java
        > 📄 Message.java
        > 📄 Sale.java
        > 📄 Type.java
        > 📄 User.java
      v ⊞ com.filter
        > 📄 EncodingFilter.java
      v ⊞ com.servlet
        > 📄 BookServlet.java
        > 📄 MessageServlet.java
        > 📄 SaleServlet.java
        > 📄 UserServlet.java
      v ⊞ com.util
        > 📄 BasicJDBC.java
  > 📂 Libraries
```

图 1-21　系统项目结构图 1

```
  > 📂 Libraries
  > 🗁 build
  v 🗁 WebContent
    > 📂 META-INF
    v 📂 WEB-INF
      > 🗁 lib
        📄 web.xml
    📄 addbook.jsp
    📄 bookmanage.jsp
    📄 failure.jsp
    📄 listbook.jsp
    📄 login.jsp
    📄 manage.jsp
    📄 salebook.jsp
    📄 salemanage.jsp
    📄 user.jsp
    📄 usercreate.jsp
    📄 usermanage.jsp
```

图 1-22　系统项目结构图 2

第 2 章　医院门诊挂号系统

2.1　需求分析

2.1.1　系统概述

医院的生存、发展在市场经济的大环境下显得至关重要。医院不仅要为患者提供服务，还要兼顾经济效益，因为这与医院的整体竞争力息息相关。各项综合因素共同决定着医院的竞争力，门诊挂号正是其中之一。到医院的门诊看病，第一件事就是挂号，挂号是医院整个就医过程的第一步。患者想要获取满意的医疗服务，必须挂一个心仪的号。

目前，许多患者到医院看病都会觉得"挂号难"。医院和患者间的交流不能对接，同时，患者就医时存在习惯性、盲目性，常导致单位时间内看病人数激增。这样极易阻塞整个看病的流程，形成我们通常所说的"挂号难"的现象。在传统的医院看病过程中，排队挂号是一个亟待解决的问题，它对医院的门诊流程运行产生巨大影响。综上所述，从医院、患者的双重角度出发，要解决看病难的问题，首先要从解决"挂号难"的问题入手。

医院预约挂号平台的设计与实现，是一个里程碑。它预示着医疗服务要翻开改革的新篇章，既能够为患者提供便捷服务，又能在很大程度上帮助医院提高服务质量。网上预约挂号平台的好处主要有以下几个方面。

（1）网络挂号快捷、方便、简单。

（2）用户可自由选择合适的专家。

（3）医院挂号的信息共享。

那么我们在设计医院门诊挂号系统时，一方面要方便患者解决挂号难、效率低等问题，另一方面要方便医院管理者更加有效地将资源分配给患者们。

2.1.2　功能需求描述

患者挂号就诊流程如下。

（1）用户注册信息。

（2）用户进入登录页面。

（3）用户进入预约页面。

（4）填写预约信息，如预约电话、就诊姓名、输入病情、医生姓名、挂号日期。

（5）用户提交后表示预约成功。

医生门诊流程如下。

（1）医生进入登录页面。

（2）医生通过查询编号进行查询。

(3) 医生执行更改、删除患者信息等操作。
(4) 进行操作后自动跳到主页面。
系统功能模块如下。
(1) 用户注册：用户分为患者和医生。患者或医生可注册用户，填写个人信息。
(2) 用户登录：通过登录页面，用户可输入用户名、密码登录系统，并跳转到主页面。
(3) 信息更改：用户登录系统后，可修改个人信息。
(4) 挂号预约：患者登录系统后，可选择医生进行挂号预约。
(5) 查询/管理患者信息：医生登录系统后，可查询、管理患者信息。
(6) 查询/管理挂号信息：医生登录系统后，可查询、管理患者挂号信息。

2.2 总体设计

2.2.1 系统总体功能结构

医院门诊挂号系统总体功能结构如图 2-1 所示。

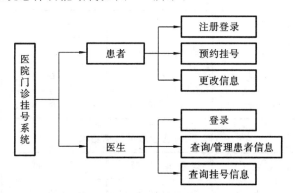

图 2-1 医院门诊挂号系统总体功能结构图

2.2.2 总体架构

医院门诊挂号系统总体架构如图 2-2 所示。

MVC 架构		
Model 层	View 层	Controller 层
DoctorDao.java	login.jsp	LoginServlet
GuahaoDao.java	doctor.jsp	DoctorServlet
UserDao.java	add.jsp	AddServlet
Doctor.java	list.jsp	ByNumberServlet
Guahao.java	patientmessage.jsp	UpdateServlet
User.java	register.jsp	
	update.jsp	

图 2-2 医院门诊挂号系统总体架构图

MVC 架构分析如下。

(1) Model(模型)层。

在 entity 和 mapper 包中,由程序 User 和 Doctor 分别包装用户和医生信息。UserDao、DoctorDao、GuahaoDao 分别定义验证用户信息的操作、医生信息的操作和挂号的操作。

(2) View(视图)层。

login、doctor 分别提供用户和医生的登录,add 和 list 分别提供用户预约和预约信息的表单。其他项则是更改、删除及查询用户信息。

(3) Controller(控制)层。

LoginServlet、DoctorServlet 分别对登入信息进行检查。AddServlet、ByNumberServlet、UpdateServlet 等对用户信息进行修改、删除、注册等操作。

2.2.3 数据库设计

患者表、医生表和挂号表的数据结构分别如表 2-1 至 2-3 所示。

表 2-1 患者表的数据结构

名 称	字段名称	数据类型	主 键	非 空
用户编号	Uid	数字(自动编号)	是	是
用户名	Uname	文本	否	是
密码	password	文本	否	是

表 2-2 医生表的数据结构

名 称	字段名称	数据类型	主 键	非 空
用户编号	Did	数字(自动编号)	是	是
用户名	Dname	文本	否	是
密码	password	文本	否	是

表 2-3 挂号表的数据结构

名 称	字段名称	数据类型	主 键	非 空
挂号编号	Gid	数字(自动编号)	是	是
预约电话	phone	数字	否	是
患者姓名	patient	文本	否	是
病情	condition	文本	否	是
医生	hName	文本	否	是
挂号日期	Gdate	日期	否	是

2.3 详细设计

2.3.1 登录功能

当医生、患者需要进入系统时,可以从登录模块输入用户名和密码进行登录。

(1) 患者登录页面文件 login.jsp 的主要代码如下:

```
<form action="./LoginServlet" method="post" onSubmit=
    "return denglu(this)" id="loginForm" name="loginForm">
<h3 style="color:black;font-size:24px;">快速登录</h3>
请输入用户名:<input type="text" name="username" placeholder=
    "请输入用户名"><br>
请输入密  码:<input type="password" name="password"
    placeholder="请输入密码"><br>
<button type="submit">登 录</button>
</form>
<a href="register.jsp"><button>去注册信息</button></a>
<a href="add.jsp"><button style="background-color:#cc99ff;margin-left:10px;
    color:black;">注册预约</button></a>
</div>
```

(2) 医生登录页面文件 doctor.jsp 的主要代码如下:

```
<form action="DoctorServlet" method="post" onSubmit="return denglu(this)"
    id="loginForm" name="loginForm">
<h3 style="color:black;font-size:24px;">医生登录</h3>
请输入用户名:<input type="text" name="doctorname" placeholder=
    "请输入用户名"><br>
请输入密  码:<input type="password" name=
    "doctorpassword" placeholder="请输入密码"><br>
<button type="submit">登 录</button>
</form>
```

医生和患者的登录页面如图 2-3 和图 2-4 所示。

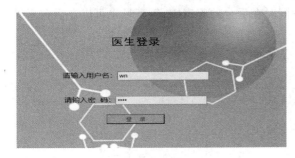

图 2-3 医生登录页面

图 2-4 患者登录页面

2.3.2 挂号预约功能

患者登录进入平台后，可以输入所要求填写的信息来进行预约。

挂号预约页面文件 add.jsp 的主要代码如下：

```html
<h3>预约信息</h3>
<form action="./AddServlet" method="post">
    预约电话:<input type="text" placeholder="请输入预约电话" name="no"></td><br>
    就诊姓名:<input type="text" placeholder="请输入就诊病人姓名" name="patient" ><br>
输入病情:<input type="text" placeholder="请输入病情" name="condation" ><br>
医生姓名:<input type="text" placeholder="请输入医生姓名" name="hName" ><br>
    挂号日期:<input type="date" style="width:140px;" placeholder="请选择挂号日期" name="createTime" ><br><br>
<button type="submit">确定预约</button><br><br>
</form>
<a href="/LoginServlet"><button>取消预约</button></a>
```

挂号预约页面如图 2-5 所示。

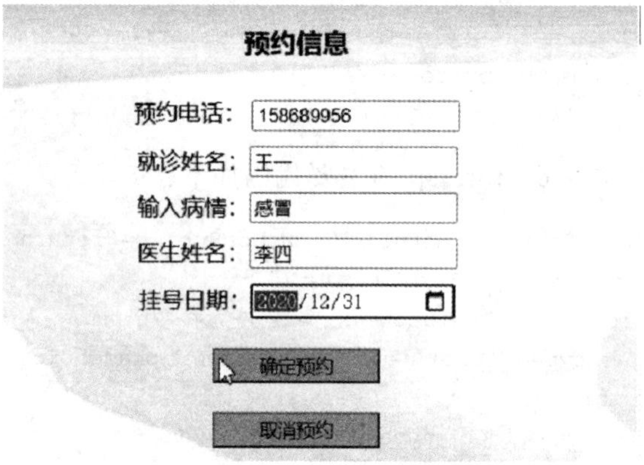

图 2-5 挂号预约页面

2.3.3 查询、更改、删除功能

医生可以登录入平台查询患者的信息；同时在后台的管理中，管理员也可以在后台对患者的信息执行查询、更改和删除操作。

(1) 医生查询页面文件 patientmessage.jsp 的主要代码如下：

```html
//查询代码
<h2 style="text-align:center;color:black;">查询用户信息</h2>
<div style="display:inline-block;">
<form action="./ByNumberServlet" style="float:left;margin-left:5px;" method="post">
<input type="text" name="no" placeholder="根据编号" style="width:170px;"/>
<input type="submit" value="查询" style="background-color:#cc99ff;margin-left:
```

```
   10px;color:black;"/>
</form>
</div>
```

(2) 医生主页面文件 list.jsp 的主要代码如下：

```html
<h2 style="text-align:center;color:black;">医院预约挂号系统</h2>
<div style="display:inline-block;">
<form action="./ByNumberServlet" style="float:left;margin-left:5px;" method=
   "post">
<input type="text" name="no" placeholder="根据编号" style="width:170px;"/>
<input type="submit" value="查询" style="background-color:#cc99ff;margin-left:
   10px;color:black;"/>
</form>
<a href="./ListServlet" style=""><button class="btn1">去主页面</button></a>
<a href="doctor.jsp" style=""><button class="btn1">退出</button></a>
</div>
<div>
<table style="margin:50px auto;" border="1px solid #ccc" cellspacing=
   "0" cellpadding="0" id="tableExcel">
<tr class="th">
<th>    预约电话      </th>
<th>    预约患者     </th>
<th>    病情     </th>
<th>    医生姓名     </th>
<th>    挂号时间     </th>
<th>  操作   </th></tr>
<c:forEach items="${list}" var="guahao">
    <tr>
        <td>${guahao.no}</td>
        <td>${guahao.patient}</td>
        <td>${guahao.condation}</td>
        <td>${guahao.hName}</td>
        <td><fmt:formatDate pattern="yyyy-MM-dd"
            value="${guahao.createTime}"/></td>
        <td style="text-align:center;">
        <a href="./QueryServlet? id=${guahao.id}" style="color:#926dde;"><button style=
            "background-color:#926dde;color:#fff;">更改</button></a>
        <a href="./DeleteServlet? id=${guahao.id}" style="color:#926dde;"><button style=
            "background-color:#f96868;;color:#fff;" onclick="return del();">删除</button></a>
        </td>
    </tr>
</c:forEach>
</table>
</div>
```

(3) 医生更改信息页面文件 update.jsp 的主要代码如下：

```
//更改代码
<h3>编辑信息</h3>
<form action="./UpdateServlet" method="post">
<input type="hidden" name="id" value="${record.id}">
预约电话:<input type="text" placeholder="请输入预约电话" value="${record.no}"
name="no"><br>
就诊姓名:<input type="text" placeholder="请输入就诊患者姓名" name="patient"
value="${record.patient}"><br>
更改病情:<input type="text" placeholder="请输入病情" value="${record.condation}"
name="condation" ><br>
医生姓名:<input type="text" placeholder="请输入医生姓名" value="${record.hName}"
name="hName"><br>
预约时间:<input type="date" style="width:140px;" value="${record.createTime}"
placeholder="请输入创建时间" name="createTime"><br>
<button type="submit">确定更改</button>
</form>
    <a href="./ListServlet"><button>取消更改</button></a>
//删除代码
<a href="./DeleteServlet?id=${guahao.id}" style="color:#926dde;"><button style=
"background-color:#f96868;;color:#fff;" onclick="return del();">删除</button></a>
```

医生查询用户信息页面如图 2-6 所示。

图 2-6　医生查询用户信息页面

医生主页面如图 2-7 所示。

图 2-7　医生主页面

医生更改患者信息页面如图 2-8 所示。

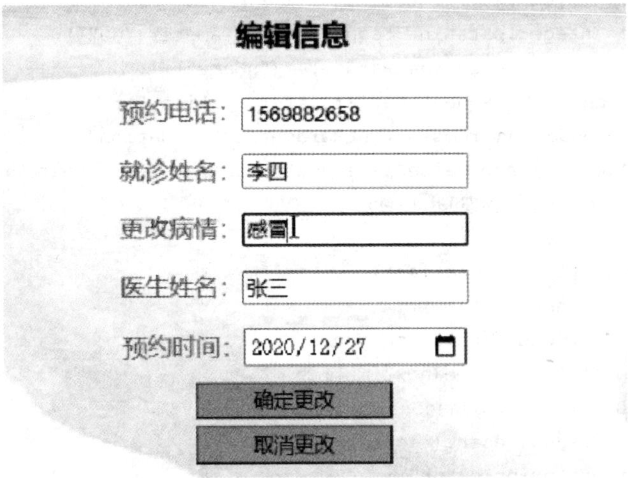

图 2-8 医生更改患者信息页面

2.4 编码实现

本节重点介绍患者挂号预约功能、医生查询功能和医生更改患者信息功能等内容。

2.4.1 患者挂号预约功能

"患者挂号预约"处理 servlet 文件 AddServlet.java 的源代码如下：

```java
public class AddServlet extends HttpServlet {
    private static final long serialVersionUID=1L;
    /**
     * @see HttpServlet#HttpServlet()
     */
    public AddServlet() {
        super();
    }
    /**
     * @see HttpServlet#doGet(HttpServletRequest request,
     *      HttpServletResponse response)
     */
    protected void doGet(HttpServletRequest request, HttpServletResponse response)
        throws ServletException, IOException {
        this.doPost(request, response);
    }
    /**
     * @see HttpServlet#doPost(HttpServletRequest request,HttpServletResponse response)
     */
    protected void doPost(HttpServletRequest request,HttpServletResponse response)
        throws ServletException, IOException {
```

```java
            request.setCharacterEncoding("utf-8");
            int no=Integer.parseInt(request.getParameter("no"));
            String hName=request.getParameter("hName");
            String patient=request.getParameter("patient");
            String condation=request.getParameter("condation");
            Date nowtime=Date.valueOf(request.getParameter("createTime"));
            Guahao record=new Guahao();
            record.setNo(no);
            record.setPatient(patient);
            record.sethName(hName);
            record.setCondation(condation);
            record.setCreateTime(nowtime);
            GuhaoDao dao=new GuhaoDao();
            boolean b=dao.insert(record);
            System.out.println("ok");
            if(b==true){
                request.getRequestDispatcher("add.jsp").forward(request,response);
            }else{
                response.sendRedirect("userlist.jsp");
            }
        }
    }
```

2.4.2 医生查询功能

"医生查询"处理 servlet 文件 ByNumberServlet.java 的源代码如下：

```java
public class ByNumberServlet extends HttpServlet {
private static final long serialVersionUID=1L;
    /**
     * @see HttpServlet#HttpServlet()
     */
    public ByNumberServlet() {
        super();
    }
    /**
     * @see HttpServlet#doGet(HttpServletRequest request,
       HttpServletResponse response)
     */
    protected void doGet(HttpServletRequest request,HttpServletResponse response)
        throws ServletException, IOException {
        this.doPost(request, response);
    }
    /**
     * @see HttpServlet#doPost(HttpServletRequest request,
       HttpServletResponse response)
     */
```

```java
    protected void doPost(HttpServletRequest request,HttpServletResponse response)
        throws ServletException, IOException {
        request.setCharacterEncoding("utf-8");
        GuhaoDao dao=new GuhaoDao();
        int no=Integer.parseInt(request.getParameter("no"));
        System.out.println(no+"");
        List<Guahao> list=dao.findByNo(no);
        request.setAttribute("list",list);
        request.getRequestDispatcher("list.jsp").forward(request, response);
    }
}
```

2.4.3 医生更改患者信息功能

"医生更改患者信息"处理 servlet 文件 UpdateServlet.java 的源代码如下：

```java
public class UpdateServlet extends HttpServlet {
    private static final long serialVersionUID=1L;
    /**
     * @see HttpServlet#HttpServlet()
     */
    public UpdateServlet() {
        super();
    }
    /**
     * @see HttpServlet#doGet(HttpServletRequest request, HttpServletResponse response)
     */
    protected void doGet(HttpServletRequest request,HttpServletResponse response)
        throws ServletException, IOException {
        this.doPost(request, response);
    }
    /**
     * @see HttpServlet#doPost(HttpServletRequest request,
       HttpServletResponse response)
     */
    protected void doPost(HttpServletRequest request,HttpServletResponse response)
        throws ServletException, IOException {
        request.setCharacterEncoding("utf-8");
        int Id=Integer.parseInt(request.getParameter("id"));
        int no=Integer.parseInt(request.getParameter("no"));
        String patient=request.getParameter("patient");
        String hName=request.getParameter("hName");
        String condation=request.getParameter("condation");
        Date nowtime=Date.valueOf(request.getParameter("createTime"));
        Guahao record=new Guahao();
        record.setNo(no);
        record.setPatient(patient);
```

```
    record.setCondation(condation);
    record.sethName(hName);
    record.setCreateTime(nowtime);
    GuhaoDao dao=new GuhaoDao();
    boolean b=dao.update(record, Id);
    System.out.println("ok");
    if(b==true){
        request.getRequestDispatcher("ListServlet").forward(request, response);
    }else{
        response.sendRedirect("list.jsp");
    }
  }
}
```

2.5 项目搭建

医院门诊挂号系统使用 eclipse 工具和 MySQL 数据库开发，系统项目结构如图 2-9、图 2-10 所示。

图 2-9 系统项目结构图 1　　　　　　　图 2-10 系统项目结构图 2

第 3 章 药品库存管理系统

3.1 需求分析

3.1.1 系统概述

随着医疗制度改革的进行,药品招标采购的逐年规范和扩大,药品管理已经成为药剂科乃至整个医院管理的重要内容。传统的管理模式已经跟不上顾客对快节奏生活的要求。随着计算机的普及和计算机软件的不断发展,越来越多的医疗机构开始重视计算机这个辅助工具。医院的售药机构需要通过计算机提高自己的工作效率、对药品实现进销存管理,以及对职工进行管理,以提高经济效益。

药品库存管理系统可提供给医疗机构的普通用户和管理者使用。

3.1.2 功能需求描述

(1)系统用户管理:允许管理员添加和删除用户,允许用户更改自己的密码,允许管理员添加或删除和修改职工信息,以及允许管理员对药品进行销售和库存管理。

(2)药品进货管理:当企业需要增加新的销售药品时,利用此模块可以添加新的药品信息、删除旧的药品信息、修改已存在的药品信息、提供所有与药品相关的各类信息、初始化库存信息、初始化供应商相关信息。

(3)药品销售管理:统一药品的销售价格,根据销售单,可以对药品的价格进行控制。

(4)库存转移管理:实际生活中容易发生药品过期和损毁,此时可以利用此模块对过期的药品和损毁的药品进行清理。

(5)职工管理:对药品企业的职工执行增加、修改、删除、查询等操作。

3.2 总体设计

3.2.1 系统总体功能结构

药品库存管理系统总体功能结构如图 3-1 所示。

管理员用例图如图 3-2 所示。

普通用户用例图如图 3-3 所示。

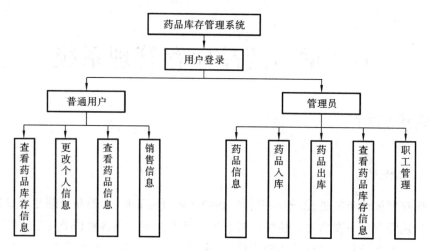

图 3-1 药品库存管理系统总体功能结构图

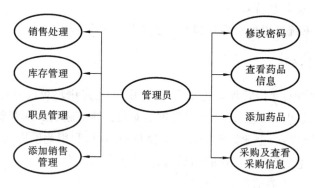

图 3-2 管理员用例图

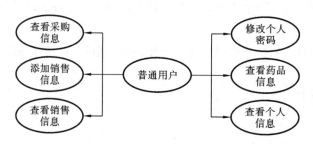

图 3-3 普通用户用例图

3.2.2 总体架构

1. 模型层

UserDao：管理员增加、删除、修改、查询用户信息，管理员或普通用户可以登录系统。

MedicineDao：管理员增加和删除药品信息，管理员或普通用户可以查看药品信息。

StockDao：管理员或普通用户可以查询库存信息。

SalesDao：管理员或普通用户可以添加销售信息。

PurchasingDao：管理员查询和增加采购信息。

2．控制层

UserServlet：查看所有员工的个人信息。

MedicineServlet：普通用户查看药品信息是跳转到普通用户药品页面，管理员查看药品信息是跳转到管理员药品页面。

StockServlet：查看库存信息是跳转到库存信息页面。

SalesServlet：查看销售信息是跳转销售信息页面。

PurchasingServlet：查看采购信息是跳转采购信息页面。

3．视图层

commonleft.jsp：显示普通用户主页面背景。

commonmain.jsp：显示普通用户主页面。

fix.jsp：显示普通用户修改个人密码页面。

Medicine.jsp：显示普通用户查看药品信息页面。

PersonalInformation.jsp：显示普通用户查看个人信息页面。

Sales.jsp：显示普通用户查看销售信息页面。

shop_add.jsp：显示普通用户添加销售信息页面。

stock.jsp：显示普通用户查看库存信息页面。

rootfix.jsp：显示管理员修改个人密码。

rootleft.jsp：显示管理员主页面背景。

rootmain.jsp：显示管理员主页面。

rootmedicine.jsp：显示管理员查看药品信息页面。

add.jsp：显示职工入职页面。

MedicineAdd.jsp：显示管理员添加药品页面。

Purchasing.jsp：显示管理员查看采购信息页面。

PurchasingAdd.jsp：显示管理员采购页面。

Sales.jsp：显示管理员查看销售信息页面。

shop_add.jsp：显示管理员添加销售信息页面。

stock.jsp：显示管理员查看库存信息页面。

Staff_Management.jsp：显示职工管理页面。

3.2.3 数据库设计

药品库存管理系统的数据库关系如图 3-4 所示。

按照系统需求，药品库存管理系统的数据库中包括用户表、药品表、药品库存表、药品采购表、药品销售表等，各表的具体定义分别如表 3-1 至表 3-5 所示。

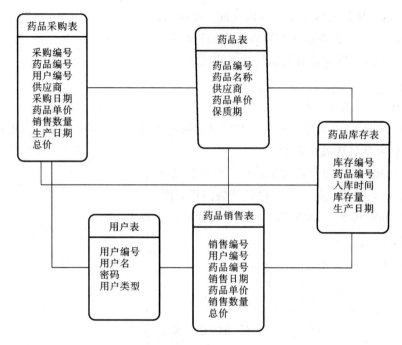

图3-4 药品库存管理系统的数据库关系图

表3-1 用户表(User)

名　　称	字段名称	数据类型	主　键	非　空
用户编号	UserID	数字(自动编号)	是	是
用户名	UserName	文本	否	是
密码	PassWord	文本	否	是
用户类型	Class	文本	否	是

说明:用户类型包括普通用户和管理员。

表3-2 药品表(Medicine)

名　　称	字段名称	数据类型	主　键	非　空
药品编号	MedicineID	数字(自动编号)	是	是
药品名称	MedicineName	文本	否	是
供应商	Suppliers	文本	否	是
药品单价	Price	数字	否	是
保质期	ShelfLife	文本	否	是

表3-3 药品库存表(Stock)

名　　称	字段名称	数据类型	主　键	非　空
库存编号	Kid	数字(自动编号)	是	是

续表

名　称	字段名称	数据类型	主　键	非　空
药品编号	MedicineID	数字	否	是
入库时间	STime	日期	否	是
库存量	Storage	数字	否	是
生产日期	ProductionDate	日期	否	是

表 3-4　药品销售表(Sales)

名　称	字段名称	数据类型	主　键	非　空
销售编号	BusinessID	数字(自动编号)	是	是
用户编号	UserID	数字	否	是
药品编号	MedicineID	数字	否	是
销售日期	SDate	日期	否	是
药品单价	Price	数值	否	是
销售数量	Quantity	数字	否	是
总价	TotalPrice	日期	否	是

表 3-5　药品采购表(Purchasing)

名　称	字段名称	数据类型	主　键	非　空
采购编号	PurchasingID	数字(自动编号)	是	是
药品编号	MedicineID	数字	否	是
用户编号	UserID	数字	否	是
供应商	Suppliers	文本	否	是
采购日期	PurchasingDate	日期	否	是
药品单价	Price	数值	否	是
销售数量	Quantity	数字	否	是
生产日期	ProductionDate	日期	否	是
总价	TotalPrice	数值	否	是

3.3　详细设计

3.3.1　药品入库

管理员进入信息管理系统页面后,点击"药品入库"按钮,出现如图 3-5 所示的库存信息页面。通过 stock.jsp 页面,调用 MedicineAdd.jsp 和 StockServlet 来控制药品入库的过程，提交之后会跳转到 stock.jsp 页面。采购信息页面如图 3-6 所示。

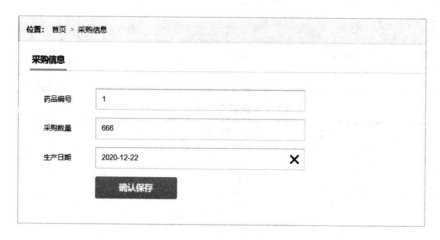

图 3-5　库存信息页面

图 3-6　采购信息页面

3.3.2　药品出库

点击药品管理下的"药品出库"（通过 Sales.jsp 来实现），弹出如图 3-7 所示的药品出库

图 3-7　药品出库信息页面

信息页面，列出出库药品的各种信息，单击"添加"链接，跳转至 shop_add.jsp 页面，弹出药品出库页面，如图 3-8 所示。通过 SalesServlet 完成药品的出库，随后跳转到销售信息的页面 sale.jsp。

图 3-8　药品出库页面

3.3.3　职工信息管理

职工信息管理是对整个系统的用户登入情况进行的管理，它通过 Staff_Management.jsp 来实现，系统通过调用 UserServlet 来控制职工添加信息的情况，通过输入用户名、用户密码和用户类型来确定是普通用户还是管理员。添加之后，数据会自动添加到 user 数据表中。管理员也可以随时删除和修改普通用户的信息。

用户信息管理页面如图 3-9 所示。

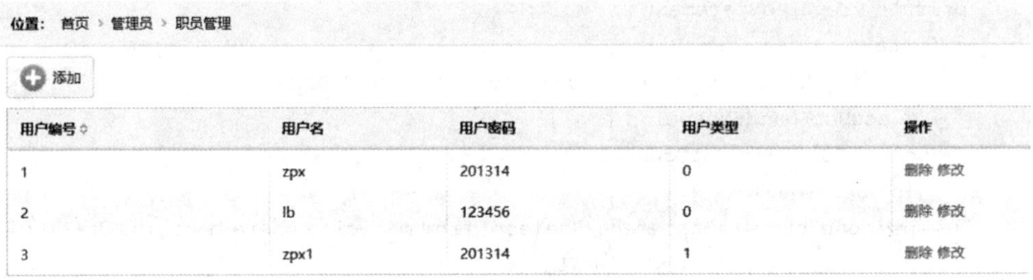

图 3-9　用户信息管理页面

添加用户页面如图 3-10 所示。

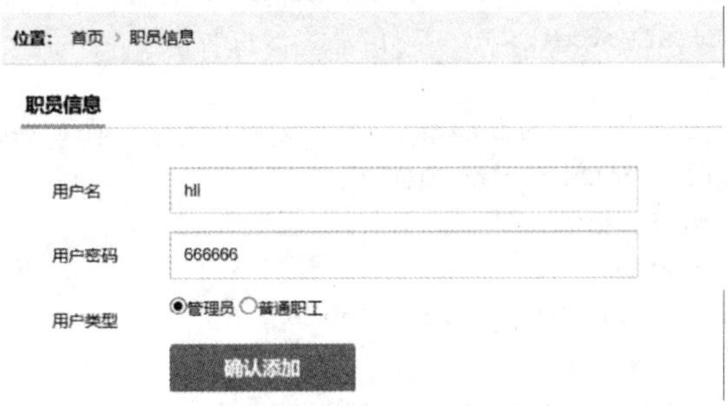

图 3-10　添加用户页面

3.4 编码实现

本节重点介绍药品入库管理和药品出库管理的相关内容。

3.4.1 药品入库管理

1. Model 层

（1）药品入库接口文件 IPurchasingDao.java 的源代码如下：

```java
//定义药品入库接口 IPurchasingDao
public interface IPurchasingDao {
    public int inserPurchasing(Purchasing p) throws SQLException,
        ClassNotFoundException;
    public List<Purchasing> selectAllPurchasing() throws SQLException,
        ClassNotFoundException;
}
```

（2）实现药品入库接口操作文件 PurchasingDaoImpl.java 的源代码如下：

```java
//定义药品入库操作类 PurchasingDaoImpl
public class PurchasingDaoImpl implements IPurchasingDao{
    private Connection conn=null;
    private PreparedStatement ps=null;
    private ResultSet rs=null;
    public int inserPurchasing(Purchasing p) throws SQLException,
        ClassNotFoundException {
        conn=DBHelper.getConnection();
        int Storage=0;
        ps=conn.prepareStatement("select* from medicine where MedicineID=?");
        ps.setInt(1, p.getMedicineID());
        rs=ps.executeQuery();
        while (rs.next()) {
            p.setPrice(rs.getDouble("Price"));
            p.setSuppliers(rs.getString("Suppliers"));
            p.setTotalPrice();
        }
        ps=conn.prepareStatement("select* from Stock where MedicineID=?");
        ps.setInt(1,p.getMedicineID());
        rs=ps.executeQuery();
        while (rs.next()) {
            Storage=rs.getInt("Storage");
        }
        ps=conn.prepareStatement("update Stock set Storage=? where MedicineID=?");
        ps.setInt(1, Storage+ p.getQuantity());
        ps.setInt(2, p.getMedicineID());
        ps.executeUpdate();
```

```java
        ps=conn.prepareStatement("insert into purchasing(MID,UID,Suppliers,
            PurchasingDate,Price,Quantity,ProductionDate,TotalPrice)
            values(?,?,?,?,?,?,?,?)");
        ps.setInt(1, p.getMedicineID());
        ps.setInt(2, p.getUserID());
        ps.setString(3, p.getSuppliers());
        ps.setString(4, p.getPurchasingDate());
        ps.setDouble(5, p.getPrice());
        ps.setInt(6, p.getQuantity());
        ps.setString(7, p.getProductionDate());
        ps.setDouble(8, p.getTotalPrice());
        int i=ps.executeUpdate();
        DBHelper.closeConntion(rs, ps, conn);
        return i;
    }
    public List<Purchasing>selectAllPurchasing() throws SQLException,
        ClassNotFoundException {
        List<Purchasing>list=new ArrayList<Purchasing>();
        conn=DBHelper.getConnection();
        ps=conn.prepareStatement("select* from Purchasing");
        rs=ps.executeQuery();
        while (rs.next()) {
            SimpleDateFormat sdf=new SimpleDateFormat("yyyy-MM-dd");
            int PurchasingID=rs.getInt("PurchasingID");
            int MedicineID=rs.getInt("MID");
            int UserID=rs.getInt("UID");
            String Suppliers=rs.getString("Suppliers");
            String PurchasingDate=sdf.format(rs.getDate("PurchasingDate"));
            double Price=rs.getDouble("Price");
            int Quantity=rs.getInt("Quantity");
            String ProductionDate=sdf.format(rs.getDate("ProductionDate"));
            double TotalPrice=rs.getDouble("TotalPrice");
            Purchasing s=new Purchasing(PurchasingID,MedicineID,UserID,Suppliers,
                PurchasingDate,Price,Quantity,ProductionDate,TotalPrice);
            list.add(s);
        }
        DBHelper.closeConntion(rs, ps, conn);
        return list;
    }
}
```

2. Controller 层

（1）药品入库处理 servlet 文件 PurchasingServlet.java 的源代码如下：

```java
public class PurchasingServlet extends HttpServlet {
    private static final long serialVersionUID=1L;
    private static IPurchasingService service=new PurchasingService();
    /**
```

```java
 * @see HttpServlet#HttpServlet()
 */
public PurchasingServlet() {
    super();
}
/**
 * @see HttpServlet#doGet(HttpServletRequest request,
   HttpServletResponse response)
 */
protected void doGet(HttpServletRequest request, HttpServletResponse response)
    throws ServletException, IOException {
    String param=request.getParameter("param");
    try {
        if ("selectAll".equals(param)) {
            selectAll(request, response);
        }else if("insert".equals(param)){
            insert(request, response);
        }
    } catch (ClassNotFoundException e) {
        e.printStackTrace();
    } catch (SQLException e) {
        e.printStackTrace();
    }
}
/**
 * @see HttpServlet#doPost(HttpServletRequest request,
   HttpServletResponse response)
 */
protected void doPost(HttpServletRequest request, HttpServletResponse response)
    throws ServletException, IOException {
    doGet(request, response);
}
private void insert(HttpServletRequest request, HttpServletResponse response)
    throws ClassNotFoundException, SQLException, ServletException, IOException {
    int MedicineID=Integer.parseInt(request.getParameter("MedicineID"));
    int UserID=Integer.parseInt(request.getParameter("UserID"));
    int Quantity=Integer.parseInt(request.getParameter("Quantity"));
    SimpleDateFormat sdf=new SimpleDateFormat("yyyy-MM-dd");   //设置日期格式
    sdf.format(new Date());                      //new Date()为获取当前系统时间
    String PurchasingDate=sdf.format(new Date());
    String ProductionDate=request.getParameter("ProductionDate");
    Purchasing s=new Purchasing(0,MedicineID,UserID,null,PurchasingDate,
        0,Quantity,ProductionDate,0);
    int i=service.inserPurchasing(s);
    if(i!=1){
        String msg="注册失败";
        System.out.print(msg);
```

```java
            request.setAttribute("msg", msg);
            request.getRequestDispatcher("register.jsp").forward(request, response);
        }else{
            selectAll(request,response);
        }
    }
    public void selectAll(HttpServletRequest request, HttpServletResponse response)
        throws ServletException, IOException, ClassNotFoundException,
            SQLException {
        //调用 service
        List<Purchasing> list=service.selectAllPurchasing();
        //显示给页面
        request.setAttribute("list", list);
        request.getRequestDispatcher("Purchasing.jsp").forward(request, response);
    }
}
```

（2）药品入库处理 service 接口文件 IPurchasingService.java 的源代码如下：

```java
public interface IPurchasingService {
    public int inserPurchasing(Purchasing p) throws SQLException,
        ClassNotFoundException;
    public List<Purchasing> selectAllPurchasing() throws SQLException,
        ClassNotFoundException;
}
```

（3）实现药品入库处理 service 接口文件 PurchasingService.java 的源代码如下：

```java
public class PurchasingService implements IPurchasingService{
    private IPurchasingDao dao= new PurchasingDaoImpl();
    @Override
    public int inserPurchasing(Purchasing p)throws SQLException,
        ClassNotFoundException{
        return dao.inserPurchasing(p);
    }
    @Override
    public List<Purchasing> selectAllPurchasing() throws SQLException,
        ClassNotFoundException{
        return dao.selectAllPurchasing();
    }
}
```

3.4.2 药品出库管理

1. Model 层

（1）定义药品出库接口文件 ISalesDao.java 的源代码如下：

```java
public interface ISalesDao {
```

```
    public int inserSales(Sales u) throws SQLException, ClassNotFoundException;
    public List<Sales> selectAllSales() throws SQLException, ClassNotFoundException;
}
```

(2) 实现药品出库接口的操作类文件 SalesDaoImpl.java 的源代码如下：

```
public class SalesDaoImpl implements ISalesDao {
    private Connection conn=null;
    private PreparedStatement ps=null;
    private ResultSet rs=null;
    public int inserSales(Sales s) throws SQLException, ClassNotFoundException {
        conn=DBHelper.getConnection();
        int Storage=0;
        double Price=0;
        ps=conn.prepareStatement("select * from medicine where MedicineID=?");
        ps.setInt(1, s.getMedicineID());
        rs=ps.executeQuery();
        while (rs.next()) {
            Price=rs.getDouble("Price");
        }
        ps=conn.prepareStatement("select * from Stock where MedicineID=?");
        ps.setInt(1,s.getMedicineID());
        rs=ps.executeQuery();
        while (rs.next()) {
          Storage=rs.getInt("Storage");
        }
        if((Storage-s.getQuantity())<0){
            return 0;
        }
        ps=conn.prepareStatement("update Stock set Storage=? where MedicineID=?");
    ps.setInt(1, Storage-s.getQuantity());
    ps.setInt(2, s.getMedicineID());
    ps.executeUpdate();
    ps=conn.prepareStatement("insert into sales(UserID,MedicineID,SDate,
        Price,Quantity,TotalPrice) values(?,?,?,?,?,?)");
    ps.setInt(1, s.getUserID());
    ps.setInt(2, s.getMedicineID());
    ps.setString(3, s.getSDate());
    ps.setDouble(4, Price);
    ps.setInt(5, s.getQuantity());
    ps.setDouble(6, Price*s.getQuantity());
    int i=ps.executeUpdate();
    DBHelper.closeConntion(rs, ps, conn);
    return i;
}
```

```java
public List<Sales> selectAllSales() throws SQLException, ClassNotFoundException {
    List<Sales> list=new ArrayList<Sales>();
    conn=DBHelper.getConnection();
    ps=conn.prepareStatement("select* from sales");
    rs=ps.executeQuery();
    while (rs.next()) {
        SimpleDateFormat sdf=new SimpleDateFormat("yyyy-MM-dd");
        int BusinessID=rs.getInt("BusinessID");
        int UserID=rs.getInt("UserID");
        int MedicineID=rs.getInt("MedicineID");
        String SDate=sdf.format(rs.getDate("SDate"));
        double Price=rs.getDouble("Price");
        int Quantity=rs.getInt("Quantity");
        double TotalPrice=rs.getDouble("TotalPrice");
        Sales s=new Sales(BusinessID,UserID,MedicineID,SDate,Price,
            Quantity,TotalPrice);
        list.add(s);
    }
    DBHelper.closeConntion(rs, ps, conn);
    return list;
}
```

2. Controller 层

（1）药品出库处理 servlet 文件 SalesServlet.java 的源代码如下：

```java
@WebServlet("/SalesServlet")
public class SalesServlet extends HttpServlet {
    private static final long serialVersionUID=1L;
    private static ISalesService service=new SalesService();
    /**
     * @see HttpServlet#HttpServlet()
     */
    public SalesServlet() {
        super();
    }
    /**
     * @see HttpServlet#doGet(HttpServletRequest request,
     HttpServletResponse response)
     */
    protected void doGet(HttpServletRequest request,HttpServletResponse response)
        throws ServletException, IOException {
        String param=request.getParameter("param");
        try {
            if ("selectAll".equals(param)) {
```

```java
                selectAll(request, response);
            } else if("insert".equals(param)){
                insert(request, response);
            }
        } catch (ClassNotFoundException e) {
            e.printStackTrace();
        } catch (SQLException e) {
            e.printStackTrace();
        }
    }
    /**
     * @see HttpServlet#doPost(HttpServletRequest request,
       HttpServletResponse response)
     */
    protected void doPost(HttpServletRequest request,HttpServletResponse response)
        throws ServletException, IOException {
        doGet(request, response);
    }
    private void insert(HttpServletRequest request, HttpServletResponse response)
        throws ClassNotFoundException, SQLException, ServletException, IOException {
        int UserID=Integer.parseInt(request.getParameter("UserID"));
        int MedicineID=Integer.parseInt(request.getParameter("MedicineID"));
        int Quantity=Integer.parseInt(request.getParameter("Quantity"));
        SimpleDateFormat sdf=new SimpleDateFormat("yyyy-MM-dd");   //设置日期格式
        sdf.format(new Date());                                    //获取当前系统时间
        String SDate=sdf.format(new Date());
        Sales s=new Sales(0,UserID,MedicineID,SDate,0,Quantity,0);
        int i=service.inserSales(s);
        if(i!=1){
            String msg="注册失败";
            System.out.print(msg);
            request.getRequestDispatcher("shop_add.jsp").forward(request, response);
        }else{
            System.out.print("插入成功");
            selectAll(request,response);
        }
    }
    public void selectAll(HttpServletRequest request, HttpServletResponse response)
        throws ServletException, IOException, ClassNotFoundException, SQLException {
        //调用service
        List<Sales> list=service.selectAllSales();
        request.setAttribute("list", list);
        request.getRequestDispatcher("Sales.jsp").forward(request, response);
    }
```

}

（2）定义药品出库处理 service 接口文件 ISalesService.java 的源代码如下：

```java
public interface ISalesService {
    public int inserSales(Sales u) throws SQLException, ClassNotFoundException;
    public List<Sales> selectAllSales() throws SQLException, ClassNotFoundException;
}
```

（3）实现药品出库处理接口文件 SalesService.java 的源代码如下：

```java
public class SalesService implements ISalesService {
    private ISalesDao dao=new SalesDaoImpl();
    public int inserSales(Sales s)throws SQLException, ClassNotFoundException{
        return dao.inserSales(s);
    }
    public List<Sales> selectAllSales() throws SQLException, ClassNotFoundException{
        return dao.selectAllSales();
    }
}
```

3.5 项目搭建

药品库存管理系统使用 eclipse 工具和 MySQL 数据库开发，系统项目结构如图 3-11 至图 3-13 所示。

图 3-11 系统项目结构图 1　　图 3-12 系统项目结构图 2　　图 3-13 系统项目结构图 3

第4章 超市商品采购管理系统

4.1 需求分析

4.1.1 系统概述

随着人们生活水平的不断提高,购物已成为一种时尚。每天都有大量的消费者在各大商场或超市中留下消费信息,因此,作为商场的管理人员,就需要有一个自动化、智能化的管理系统处理这些信息。在传统的手工管理中,往往采用人工清点的方式来掌握超市中现有的商品,使用手工记账的方式来掌握商品的进货和销售情况。每天借助人工来记录纷繁复杂的进货和销售情况,工作量将非常巨大。基于这种情况,我们使用所学的知识,开发了一个既能节约资金又能完成超市日常管理任务的系统。

4.1.2 功能需求描述

考虑到系统的安全方面,我们规定使用不同权限的登录用户,如营业员、采购员、库存管理员和经理,每个用户都有自己的功能,各司其职。

其中各用户的主要功能如下。
- 营业员处理收银、找零。
- 采购员进行商品采购管理和供货商管理。
- 库存管理员对商品进行入库管理和库存管理。
- 经理对超市所有人员进行账户管理、商品销售情况的管理、销售额的统计。

1. 收银功能

营业员根据客户购物车内所选的商品来结账,收到钱款找零并记录。

2. 商品管理功能

采购员:将采购的商品添加到采购表中,并管理供货商表的信息,包括添加和删除商品的信息。在采购表中,可以选择采购商品的数量来进货。在供货商表中,采购员可以查看或删除表中的信息,可以手动添加输入供货商的名称、电话、商品类型,也可以一键重置将列表清零。

库存管理员:将采购表中的商品转入库存表中,并管理库存表中商品的信息。库存管理员可以查看或删除库存表中的信息,也可以将库存表中的商品添加到商品表中,顾客在购物的时候可以直接看到商品表中的商品。

3. 统计功能

经理管理超市所有人员的账号信息,可以添加和删除用户账号,可以查看所有的现存账号,可以手动输入用户名和密码建立账户,也可以重置所有账号信息,还可以一键查询卖出去的商品和总的销售额。

4.2 总体设计

4.2.1 总体功能结构

超市商品采购管理系统总体功能结构如图 4-1 所示。

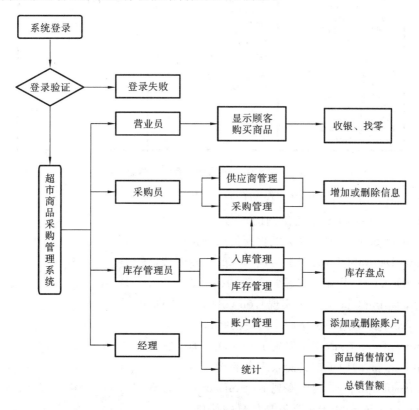

图 4-1 超市商品采购管理系统总体功能结构图

4.2.2 总体架构

超市商品采购管理系统总体架构如图 4-2 所示。

1. Model 层

BasicJDBC.java：为整个应用程序目标数据库提供一个统一的连接对象。
EncodingFilter.java：为字符编码过滤器。
User.java：用于封装用户的信息。
Product.java：用于封装商品的信息。
Kucun.java：用于封装库存的信息。
Jinhuo.java：用于封装进货的信息。

MVC架构		
Model层	View层	Controller层
GonghuoDao.java GonghuoDaoImpl.java GouwuDao.java GouwuDaoImpl.java JinhuoDao.java JinhuoDaoImpl.java KucunDao.java KucunDaoImpl.java ProductDao.java ProductDaoImpl.java UserDao.java UserDaoImpl.java Gonghuo.java Gouwu.java Jinhuo.java Kucun.java Product.java User.java EncodingFilter.java BasicJDBC.java	采购员.jsp 经理.jsp 库存管理员.jsp 营业员.jsp adduser.jsp failure.jsp gonghuoshang.jsp listgonghuo.jsp listgouwu.jsp listjinhuo.jsp listkucun.jsp listkucun2.jsp listuser.jsp listxiaoshou.jsp login.jsp	gonghuoServlet.java gouwuServlet.java jinhuoServlet.java kucunServlet.java loginServlet.java productServlet.java userServlet.java

图 4-2 超市商品采购管理系统总体架构图

Gouwu.java：用于封装用户选择的信息。

Gonghuo.java：用于封装进货商的信息。

GonghuoDao.java：用于定义供货商信息的接口。

GonghuoDaoImpl.java：用于实现 GonghuoDao 接口的类。

GouwuDao.java：用于定义用户购物的接口。

GouwuDaoImpl.java：用于实现 GouwuDao 接口的类。

JinhuoDao.java：用于定义商家进货的接口。

JinhuoDaoImpl.java：用于实现 JinhuoDao 接口的类。

KucunDao.java：用于定义商家库存的接口。

KucunDaoImpl.java：用于实现 KucunDao 接口的类。

ProductDao.java：用于定义商品信息的接口。

ProductDaoImpl.java：用于实现 Product 接口的类。

UserDao.java：用于定义用户信息的接口。

UserDaoImpl.java：用于实现 UserDao 接口的类。

2. View 层

采购员.jsp：提供商家选择商品的页面。

经理.jsp：提供管理员管理的页面。

库存管理员.jsp：提供查看库存的页面。

营业员.jsp：提供收银的页面。

adduser.jsp：提供管理员增添用户的页面。

failure.jsp：对用户的异常操作进行报错。

gonghuoshang.jsp：提供添加供货商信息显示的页面。

listgonghuo.jsp：提供显示供货商信息列表的页面。
listgouwu.jsp：提供显示用户购买商品的页面。
listjinhuo.jsp：提供超市进货商品的显示页面。
listkucun.jsp：提供超市库存信息的显示页面。
listuser.jsp：提供用户信息的显示页面。
listxiaoshou.jsp：提供销售员对具体商品收银的页面。
login.jsp：提供用户登录的表单，可以输入用户名和密码。

3. Controller 层

gonghuoServlet.java：接收请求的参数。首先使用 Gonghuo 类的构造方法实例化对象，再调用 Gonghuo 类的 getXXX() 方法得到每个变量的值。其次通过 GonghuoDaoImpl 类无参的构造方法实例化 GonghuoDao 类的对象。

gouwuServlet.java：接收请求的参数。首先使用 Gouwu 类的构造方法实例化对象，再调用 Gouwu 类的 getXXX() 方法得到每个变量的值。其次通过 GouwuDaoImpl 类无参的构造方法实例化 GouwuDao 类的对象。

jinhuoServlet.java：接收请求的参数。首先使用 Jinhuo 类的构造方法实例化对象，再调用 Jinhuo 类的 getXXX() 方法得到每个变量的值。其次通过 JinhuoDaoImpl 类无参的构造方法实例化 JinhuoDao 类的对象。

kucunServlet.java：接收请求的参数。首先使用 Kucun 类的构造方法实例化对象，再调用 Kucun 类的 getXXX() 方法得到每个变量的值。其次通过 KucunDaoImpl 类无参的构造方法实例化 KucunDao 类的对象。

loginServlet.java：接收请求的参数。首先使用 User 类的构造方法实例化对象，再调用 User 类的 getXXX() 方法得到每个变量的值。其次通过 UserDaoImpl 类无参的构造方法实例化 UserDao 类的对象。

productServlet.java：接收请求的参数。首先使用 Product 类的构造方法实例化对象，再调用 Product 类的 getXXX() 方法得到每个变量的值。其次通过 ProductDaoImpl 类无参的构造方法实例化 ProductDao 类的对象。

userServlet.java：接收请求的参数。首先使用 User 类的构造方法实例化对象，再调用 User 类的 getXXX() 方法得到每个变量的值。其次通过 UserDaoImpl 类无参的构造方法实例化 UserDao 类的对象。

4.2.3 数据库设计

按照超市商品采购管理系统的需求，数据库中包括用户信息表、供货商表、商品表、进货表、购物表等。表的具体定义如表 4-1 至表 4-5 所示。

表 4-1 用户信息表

名 称	字段名称	数据类型	主 键	非 空
用户 ID	Uid	数字(自动编号)	是	是
用户名	Uname	文本	否	是

续表

名　　称	字段名称	数据类型	主　键	非　空
密码	Password	文本	否	是
用户类型	Utype	文本	否	是
描述	Udesc	文本	否	否

说明：用户类型包括客户、采购员、营业员、库存管理员、管理员（经理）。

表 4-2　供货商表

名　　称	字段名称	数据类型	主　键	非　空
供货商 ID	Gid	数字（自动编号）	是	是
供货商名称	Gname	文本	否	是
供货商类型	Gtype	文本	否	是
电话	Phone	文本	否	是

表 4-3　商品表

名　　称	字段名称	数据类型	主　键	非　空
商品 ID	Sid	数字（自动编号）	是	是
商品名称	Sname	文本	否	是
商品类型	Stype	文本	否	是
单价	Price	数字	否	是
库存	Skucun	数字	否	是

表 4-4　进货表

名　　称	字段名称	数据类型	主　键	非　空
进货 ID	Jid	数字（自动编号）	是	是
商品 ID	Sid	数字	否	是
数量	Count	数字	否	是
供货商 ID	Gid	数字	否	是
采购员	Uid	数字	否	是
进货日期	Jdate	日期	否	是

表 4-5 购物表

名 称	字段名称	数据类型	主 键	非 空
购物 ID	Wid	数字(自动编号)	是	是
商品 ID	Sid	数字	否	是
商品名称	Sname	文本	否	是
商品单价	Price	数字	否	是
数量	Count	数字	否	是
金额	Jine	数字	否	是
客户 ID	Uid	数字	否	是
营业员 ID	Yid	数字	否	是
购买时间	Wdate	日期	否	是

4.3 详细设计

4.3.1 采购员

采购员主要是实现商品的采购功能,并将数据储存至数据库中。在商品采购页面下还需要实现对供货商信息的管理,以及显示除供货商信息的列表。商品采购页面如图 4-3 所示,添加供货商页面如图 4-4 所示,供货商信息页面如图 4-5 所示,采购顺序图如图 4-6 所示。

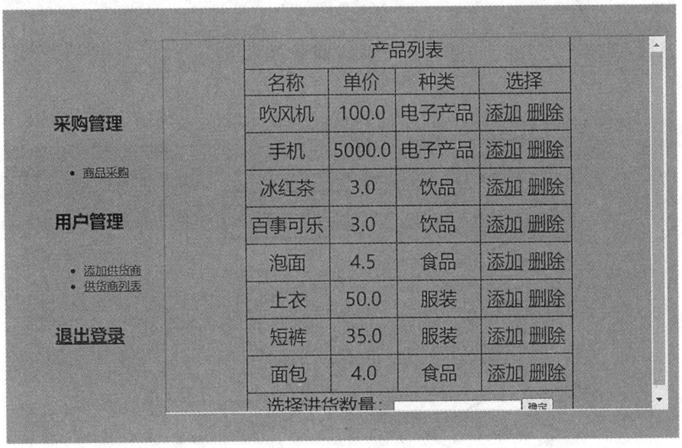

图 4-3 商品采购页面

采购功能:进货表中包含有数据,采购员在这里可以采购商品。采购员浏览商品,如果想进货某一种商品,就要在下方的输入框中输入进货的数量,然后点击商品后对应的"添加"按钮,这样就会将商品的信息和刚才的数量传递给 jinhuoServlet 并调用 add() 方法,在 add()

图 4-4　添加供货商页面

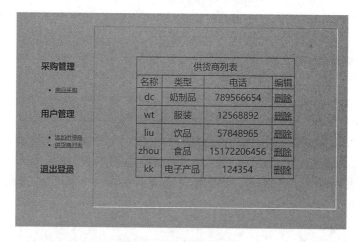

图 4-5　供货商信息页面

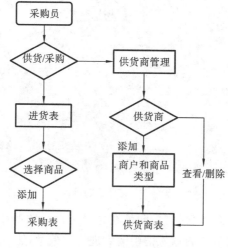

图 4-6　采购顺序图

方法中调用JinhuoDaoImpl类里的addjinhuo()方法,就会将商品添加到采购表中。

供货商管理功能:"添加供货商"链接中用到了表单的提交方式,包含三个输入框参数,分别是名称、类型和电话。输入完后点击"添加"按钮,会将这三个参数传送到gonghuoServlet中。调用add()方法,在该方法中调用GonghuoDaoImpl类的addgonghuo()方法,将数据插入供货商表中。查看供货商,调用gonghuoServlet中的list()方法,在该方法中调用GonghuoDaoImpl类里的getlist()方法,再将数据添加到数组中。然后将这个数组传送到listgonghuo.jsp显示页面。这个页面提供删除功能,在对应的供货商列表后面点击"删除"按钮,就会把这一行的各字段的值作为参数传递给gonghuoServlet,调用del()方法。在del()方法中调用GonghuoDaoImpl类的delgonghuo()方法,就会在供货商表中删除该供货商。

4.3.2 超市库存管理员

超市库存管理员需要从数据库中读取已采购的信息,并在操作过程中将其添加至库存中,还需提供库存货品的显示页面。商品入库页面如图4-7所示,查看库存页面如图4-8所示,库存管理顺序图如图4-9所示。

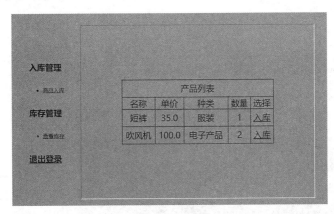

图4-7 商品入库页面

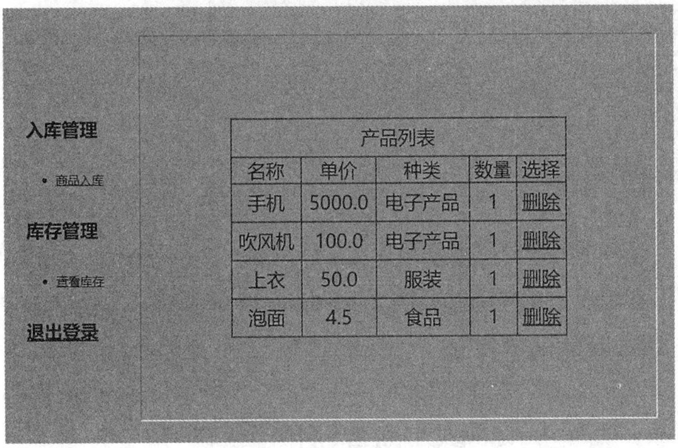

图4-8 查看库存页面

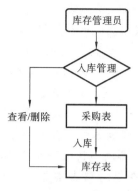

图 4-9 库存管理顺序图

入库管理:是将采购员采购在采购表里的东西进行入库。首先调用 kucunServlet 中的 listcaigou() 方法,在该方法中调用 KucunDaoImpl 类的 getlistcaigou() 方法,并返回采购表中数据的数组,然后返回 listkucun.jsp 页面。在这个页面中点击对应商品后的"入库",就会将字段的值作为参数传送到 kucunServlet 中,再调用 add() 方法,在该方法中调用 KucunDaoImpl 类的 addkucun() 方法,将数据插入库存表中。同时调用 ProductDaoImpl 类中的 addProduct() 方法,将数据插入商品表中。然后调用 JinhuoDaoImpl 类中的 deljinhuo() 方法,在采购表中删除已经入库的商品。

库存管理:调用 kucunServlet 中的 listkucun() 方法,在该方法中调用 KucunDaoImpl 类的 getListkucun() 方法,并返回库存表中数据的数组。然后返回 listkucun2.jsp 页面。在这个页面中可以查看库存表里的东西。点击对应商品后面的"删除"按钮,就会将这行字段的值传送给 kucunServlet,调用 del() 方法,再在该方法中调用 KucunDaoImpl 里的 delkucun() 方法,以及调用 ProductDaoImpl 类的 delProduct() 方法,就会删除对应行的数据。

4.3.3 营业员

营业员需要读取库存中的信息,然后通过选择商品,输入对应的金额,并返回收银找零页面。

收银找零页面如图 4-10 所示,收银顺序图如图 4-11 所示。

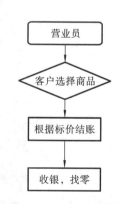

图 4-10 收银找零页面   图 4-11 收银顺序图

首先,营业员根据客户要购买的东西,在库存表里选择商品。点击"添加"按钮,表示客户选购了这个商品。完成商品添加后,会将参数传送到 gouwuServlet,调用 add() 方法,在该方法中调用 GouwuDaoImpl 类的 addgouwu(),这个商品就会添加到购物表中,然后在购物表中显示这个商品。如果在购物表中点击"删除"按钮,则会将参数传送到 gouwuServlet,调用 del() 方法,在该方法中调用 GouwuDaoImpl 类的 delgouwu(),这个商品就会从购物表中删除，表示客户不选择这个商品。在每次点击“添加”或“删除”按钮后，调用 GouwuDaoImpl 类的 listgouwu()方法，会让表格中的数据实时地更新。在 gouwuServlet 中，float 类的变量用来存放根据单价和数量计算出的单个商品的总价。在 gouwuServlet 层里调用 GouwuDaoImpl 类的 jiesuan()方法，当点击“添加”或“删除”按钮时都会调用这个方法，这样计算出来的总价在〈span〉标签中就会实时地变动。

收银功能：这里设置一个输入框，用来表示用户支付的实际价格。计算原理是，在 jsp 中编写一个 js，根据 id 来获取输入框和代表总价的〈span〉里的值进行计算，然后赋值给另一个代表找零的〈span〉里。我们再设置一个〈button〉，点击“结账”按钮就会调用 js 中的方法计算出价格。

结账完成后要点击清空购物表，会传送一个 action 参数中的值到 gouwuServlet，调用 delall()方法，在该方法中调用 GouwuDaoImpl 类的 delgouwuall()方法清空购物表。

4.3.4　管理员（经理）

管理员可以显示销售情况，可以管理用户账号，可以执行账户的增、删、改、查操作，还可以显示用户列表。销售情况页面如图 4-12 所示，添加用户页面如图 4-13 所示，查看用户页面如图 4-14 所示，管理员顺序图如图 4-15 所示。

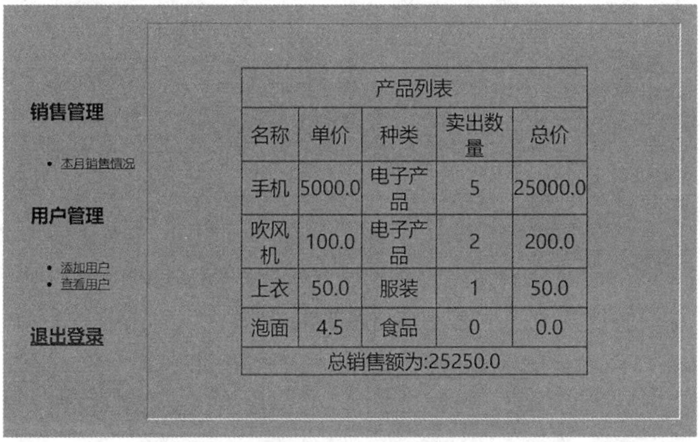

产品列表				
名称	单价	种类	卖出数量	总价
手机	5000.0	电子产品	5	25000.0
吹风机	100.0	电子产品	2	200.0
上衣	50.0	服装	1	50.0
泡面	4.5	食品	0	0.0
总销售额为:25250.0				

图 4-12　销售情况页面

人员账户管理：点击“用户管理”进入 adduser.jsp 页面，这个页面包含两个输入框，一个是账号，一个是密码。单击“添加”按钮，会将数据提交到 userServlet 中。调用 add()方法，在该方法中调用 UserDaoImpl 里的 addUser()方法，会将数据添加到 user 表中。

删除账号要在 listuser.jsp 页面进行。点击“删除”按钮，会将字段的值作为参数传送给 userServlet，然后调用 del()方法，在该方法中调用 UserDaoImpl 里的 delUser()方法。这样就可以在 user 表中删除数据。

销售情况要在商品表中查看，该表中包含每种商品卖出的数量，即单个商品卖出的总价。计算总销售额是调用 productServlet 中的 jiesuan()方法，在该方法中调用 ProductDaoImpl 里的 jiesuan()方法，然后将这个值传送到 listxiaoshou.jsp 页面，就可以看到总销售额。

图 4-13　添加用户页面

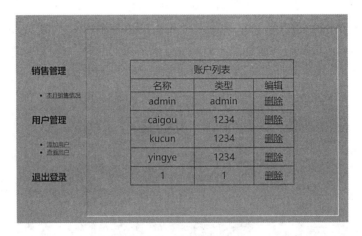

图 4-14　查看用户页面

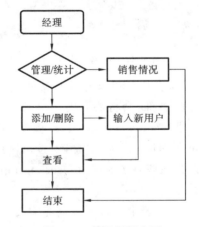

图 4-15　管理员顺序图

4.4 编码实现

本节重点介绍采购管理、供应商管理、购物缴费、产品列表等模块。公共模块、用户登录、库存管理、用户管理等模块请参见其他章节。

4.4.1 采购管理

点击"商品采购",向 jinhuoServlet 层传送一个 action 参数,根据该参数的值 list 调用 jinhuoServlet 层的 list()方法。在 list()方法中,通过调用 JinhuoDaoImpl 类的无参构造方法实例化 JinhuoDao 类的对象,再通过调用这个对象查询进货表中的方法,这个方法返回 jinhuoServlet 层的一个数组,jinhuoServlet 层将这个数组转发给相应的 jsp 页面,该页面就包含了进货表中的数据。点击"添加"按钮,向 jinhuoServlet 层传送一个 action 参数,根据该参数的值 add 将表中每个字段的值作为参数。jinhuoServlet 层根据 action 参数的值选择 add()方法。在 add()方法里接收传送来的参数,通过调用 Jinhuo 类的构造方法实例化这个类的对象,再通过这个对象调用添加到采购表中的方法。这样就实现了采购管理的功能。

1. Model 层

(1) 定义采购类接口 JinhuoDao 文件 JinhuoDao.java,源代码如下:

```
public interface JinhuoDao {
    public List<Jinhuo> getList();
    public int addjinhuo (Jinhuo j);
    public int deljinhuo (Jinhuo j);
}
```

(2) 定义采购操作类 JinhuoDaoImpl 文件 JinhuoDaoImpl.java,源代码如下:

```
public class JinhuoDaoImpl implements JinhuoDao{
    private BasicJDBC db=null;
    private Connection con=null;
    public JinhuoDaoImpl(){
        db=new BasicJDBC();
        con=db.getCon();
    }
//增加采购信息
    public int addjinhuo(Jinhuo j){
        PreparedStatement ps=null;
        String sql="insert into caigou(name,price,type,num) values(?,?,?,?)";
        int n=0;
        try{
            ps=con.prepareStatement(sql);
            ps.setString(1,j.getname());
            ps.setFloat(2,j.getprice());
            ps.setString(3,j.gettype());
```

```java
                ps.setInt(4,j.getnum());
                n=ps.executeUpdate();
                System.out.println(n);
            }catch (SQLException e){
                e.printStackTrace();
            }finally{
                db.closeAll(con,ps,null);
            }
            return n;
        }
        //删除采购信息
        public int deljinhuo(Jinhuo j){
            PreparedStatement ps=null;
            String sql="delete from caigou where name=? and price=? and type=?";
            int n=0;
            try{
                ps=con.prepareStatement(sql);
                ps.setString(1,j.getname());
                ps.setFloat(2,j.getprice());
                ps.setString(3,j.gettype());
                n=ps.executeUpdate();
                System.out.println(n);
            } catch (SQLException e){
                e.printStackTrace();
            } finally{
                db.closeAll(con,ps,null);
            }
            return n;
        }
        @Override
        //获取采购信息列表
        public List<Jinhuo>getList(){
            PreparedStatement ps=null;
            String sql="select* from jinhuo";
            ResultSet rs=null;
            List<Jinhuo>list=new ArrayList<Jinhuo>();
            try{
                ps=con.prepareStatement(sql);
                rs=ps.executeQuery();
                while(rs.next()){
                    Jinhuo j=new Jinhuo(rs.getString(1),rs.getFloat(2),rs.getString(3));
                    list.add(j);
                }
            }catch (SQLException e){
```

```
            e.printStackTrace();
        }finally{
            db.closeAll(con,ps,rs);
        }
        return list;
    }
}
```

2. View 层

提供商家选择商品的页面文件,采购员.jsp 的源代码如下:

```jsp
<%@page language="java" contentType="text/html; charset=utf-8"
    pageEncoding="utf-8"%>
<!DOCTYPE html>
<html>
<head>
<meta charset="utf-8">
<title>Insert title here</title>
</head>
<body>
<style type="text/css">
body{
    background-image:url("背景.jpg");
    background-size:100% 100%;
    background-attachment:fixed;
}
.text {
    width:166px;
}
div {
    position: absolute;
    top: 50%;
    left: 50%;
    -webkit-transform: translate(-50%,-50%);
    -moz-transform: translate(-50%,-50%);
    -ms-transform: translate(-50%,-50%);
    -o-transform: translate(-50%,-50%);
    transform: translate(-50%,-50%);
}
</style>
</head>
<body>
<div>
<table style="width:850px;">
<tr>
```

```html
<td>
    <table style="width:150px;">
    <tr>
        <td><h2>采购管理</h2></td>
    </tr>
    <tr>
        <td>
            <ul>
                <li>
                    <a href="jinhuoServlet?action=list" target="test">商品采购</a>
                </li>
            </ul>
        </td>
    </tr>
    <tr>
        <td><h2>用户管理</h2></td>
    </tr>
    <tr>
        <td>
            <ul>
                <li>
                    <a href="gonghuoshang.jsp" target="test">添加供货商</a>
                </li>
                <li>
                    <a href="gonghuoServlet? action=list" target="test">供货商列表</a>
                </li>
            </ul>
        </td>
    </tr>
    <tr>
    <td><a href="login.jsp"><h2>退出登录</h2></a></td>
    </tr>
    </table>
</td>
<td rowspan="3" style="width:700px;height:500px">
<iframe src="" name="test" width="100%" height="100%"></iframe>
</tr>
</table>
</div>
</body>
</html>
```

3. Controller 层

采购处理 servlet 文件 jinhuoServlet.java 的主要代码如下：

```java
@WebServlet("/jinhuoServlet")
public class jinhuoServlet extends HttpServlet {
    private static final long serialVersionUID=1L;
    public jinhuoServlet() {
        super();
    }
    protected void doGet(HttpServletRequest request, HttpServletResponse response)
        throws ServletException, IOException {
        doPost(request, response);
    }
    protected void doPost(HttpServletRequest request, HttpServletResponse response)
        throws ServletException, IOException {
        String action=request.getParameter("action");
        if(action.equals("list"))
            list(request,response);
        if(action.equals("del"))
            del(request,response);
        if(action.equals("add"))
            add(request,response);
    }
    protected void add(HttpServletRequest request,HttpServletResponse response)
            throws ServletException,IOException{
        String name=request.getParameter("name");
        String price=request.getParameter("price");
        String type=request.getParameter("type");
        String num=request.getParameter("num");
        System.out.println(name);
        System.out.println(price);
        System.out.println(type);
        Jinhuo jinhuo=new Jinhuo(name,Float.parseFloat(price),
            type,Integer.valueOf(num));
        //创建功能类对象来实现功能
        JinhuoDao jdao=new JinhuoDaoImpl();
        try{
            jdao.addjinhuo(jinhuo);
        }
        catch(Exception e){
            e.printStackTrace();
            request.getRequestDispatcher("/failure.jsp").forward(request,response);
            return;
        }
        response.sendRedirect("listjinhuo.jsp");
    }
    protected void del(HttpServletRequest request,HttpServletResponse response)
```

```java
        throws ServletException,IOException{
        String name=request.getParameter("name");
        String price=request.getParameter("price");
        String type=request.getParameter("type");
        String num=request.getParameter("num");
        System.out.println(name);
            System.out.println(price);
            System.out.println(type);
        Jinhuo jinhuo=new Jinhuo(name,Float.parseFloat(price),
            type,Integer.valueOf(num));
        //创建功能类对象来实现功能
        JinhuoDao jdao=new JinhuoDaoImpl();
        try{
            jdao.deljinhuo(jinhuo);
        }
        catch(Exception e){
            e.printStackTrace();
            request.getRequestDispatcher("/failure.jsp").forward(request,response);
            return;
        }
        response.sendRedirect("listjinhuo.jsp");
    }
    protected void list (HttpServletRequest request,HttpServletResponse response)
        throws ServletException,IOException{
        {
            JinhuoDao jdao=new JinhuoDaoImpl();
            List<Jinhuo>jinhuolist=jdao.getList();
            HttpSession session=request.getSession();
            session.setAttribute("jinhuolist",jinhuolist);
            response.sendRedirect("listjinhuo.jsp");
        }
    }
}
```

4.4.2 供货商管理

点击"添加供货商"进入 jsp 页面,该页面包含三个输入框,分别用于填写供货商名称、供货类型和联系方式。填写完成后点击"添加"按钮,会通过表单的方式提交到gonghuoServlet,根据 action 参数的值 add 调用 gonghuoServlet 层的 add()方法。在 add()方法中接收传送过来的参数,通过调用 Gonghuo 类的有参构造方法实例化 Gonghuo 类的对象,通过调用 GonghuoDaoImpl 类的无参构造方法实例化 GonghuoDao 类的对象,然后调用这个对象的添加到供货商表中的方法,以 Gonghuo 类的对象作为参数。

1. Model 层

(1)定义供货商信息接口文件 GonghuoDao.java,源代码如下:

```java
public interface GonghuoDao{
    public List<Gonghuo> getList();
    public int addgonghuo(Gonghuo p);
    public int delgonghuo(Gonghuo p);
}
```

(2) 实现 GonghuoDao 接口文件 GonghuoDaoImpl.java,源代码如下:

```java
package com.lx.dao;
import java.sql.Connection;
import java.sql.PreparedStatement;
import java.sql.ResultSet;
import java.sql.SQLException;
import java.util.ArrayList;
import java.util.List;
import com.lx.entity.Gonghuo;
import com.lx.util.BasicJDBC;
public class GonghuoDaoImpl implements GonghuoDao{
    private BasicJDBC db=null;
    private Connection con=null;
    public GonghuoDaoImpl(){
        db=new BasicJDBC();
        con=db.getCon();
    }
    public int addgonghuo(Gonghuo p){
        PreparedStatement ps=null;
        String sql="insert into gonghuo(name,type,Telephone) values(?,?,?)";
        int n=0;
        try{
            ps=con.prepareStatement(sql);
            ps.setString(1,p.getname());
            ps.setString(2,p.gettype());
            ps.setString(3,p.gettelephone());
            n=ps.executeUpdate();
            System.out.println(n);
        }catch (SQLException e){
            e.printStackTrace();
        }finally{
            db.closeAll(con,ps,null);
        }
        return n;
    }
    public int delgonghuo(Gonghuo p){
        PreparedStatement ps=null;
        String sql="delete from gonghuo where name=? and type=?";
```

```java
        int n=0;
        try{
            ps=con.prepareStatement(sql);
            ps.setString(1,p.getname());
            ps.setString(2,p.gettype());
            n=ps.executeUpdate();
            System.out.println(n);
        }catch (SQLException e){
            e.printStackTrace();
        }finally{
            db.closeAll(con,ps,null);
        }
        return n;
    }
    public List<Gonghuo> getList(){
        PreparedStatement ps=null;
        String sql="select *  from gonghuo";
        ResultSet rs=null;
        List<Gonghuo> list=new ArrayList<Gonghuo>();
        try{
            ps=con.prepareStatement(sql);
            rs=ps.executeQuery();
            while(rs.next()){
                Gonghuo p=new Gonghuo(rs.getString(1),rs.getString(2),rs.getString(3));
                list.add(p);
            }
        }catch (SQLException e){
            e.printStackTrace();
        }finally{
            db.closeAll(con,ps,rs);
        }
        return list;
    }
}
```

2. View层

添加供货商信息显示页面文件Gonghuoshang.jsp，源代码如下：

```jsp
<%@page language="java" contentType="text/html; charset=utf-8"
    pageEncoding="utf-8"%>
<!DOCTYPE html>
<html>
<head>
<meta charset="utf-8">
<title>Insert title here</title>
```

```html
<style type="text/css">
body{
    background-size:100% 100%;
    background-attachment:fixed;
    }
.text{
        width:166px;
}
table{
font-size:25px;
}
div{
    position: absolute;
    top: 50%;
    left: 50%;
    border:1px solid black;
    -webkit-transform: translate(-50%, -50%);
    -moz-transform: translate(-50%, -50%);
    -ms-transform: translate(-50%, -50%);
    -o-transform: translate(-50%, -50%);
    transform: translate(-50%, -50%);
}
</style>
</head>
<body>
<div>
<form method="post">
<table>
<tr><th colspan="2">处理供货商</th></tr>
<tr><td>名称:</td><td><input type="text" name="name" class="text"></td></tr>
<tr><td>类型:</td><td><input type="text" name="type" class="text"></td></tr>
<tr><td>电话:</td><td><input type="text" name="telephone" class="text"></td></tr>
<tr><td colspan="2" align="center">
<input type="reset" value="重置">
<input type="submit" value="添加" formaction="gonghuoServlet?action=add">
</td></tr>
</table>
</form>
</div>
</body>
</html>
```

3. Controller 层

处理供货商 servlet 文件 gonghuoServlet.java,源代码如下:

```java
package com.lx.servlet;
import java.io.IOException;
import java.util.List;
import javax.servlet.ServletException;
import javax.servlet.annotation.WebServlet;
import javax.servlet.http.HttpServlet;
import javax.servlet.http.HttpServletRequest;
import javax.servlet.http.HttpServletResponse;
import javax.servlet.http.HttpSession;
import com.lx.dao.GonghuoDao;
import com.lx.dao.GonghuoDaoImpl;
import com.lx.entity.Gonghuo;
@WebServlet("/gonghuoServlet")
public class gonghuoServlet extends HttpServlet {
    private static final long serialVersionUID=1L;
    public gonghuoServlet() {
        super();
    }
    protected void doGet(HttpServletRequest request, HttpServletResponse response)
        throws ServletException, IOException {
        doPost(request,response);
    }
    protected void doPost(HttpServletRequest request, HttpServletResponse response)
        throws ServletException, IOException {
        String action=request.getParameter("action");
        if(action.equals("add"))
            add(request,response);
        if(action.equals("list"))
            list(request,response);
        if(action.equals("del"))
            {del(request,response);
            list(request,response);}
    }
    protected void add(HttpServletRequest request,HttpServletResponse response)
            throws ServletException,IOException{
            String name=request.getParameter("name");
            String type=request.getParameter("type");
            String telephone=request.getParameter("telephone");
            Gonghuo gonghuo=new Gonghuo(name,type,telephone);
            //创建功能类对象来实现功能
            GonghuoDao gdao=new GonghuoDaoImpl();
            int n=0;
            try{
                n=gdao.addgonghuo(gonghuo);
```

```
        }
        catch(Exception e){
            e.printStackTrace();
            request.getRequestDispatcher("/failure.jsp").forward(request,response);
            return;
        }
        response.sendRedirect("gonghuoshang.jsp");
    }
    protected void del(HttpServletRequest request,HttpServletResponse response)
            throws ServletException,IOException{
        String name=request.getParameter("name");
        String type=request.getParameter("type");
        String telephone=request.getParameter("telephone");
        Gonghuo gonghuo=new Gonghuo(name,type,telephone);
        //创建功能类对象来实现功能
        GonghuoDao gdao=new GonghuoDaoImpl();
        int n=0;
        try{
            n=gdao.delgonghuo(gonghuo);
        }
        catch(Exception e){
            e.printStackTrace();
            request.getRequestDispatcher("/failure.jsp").forward(request,response);
            return;
        }
    }
    protected void list (HttpServletRequest request,HttpServletResponse response)
            throws ServletException,IOException{
        GonghuoDao gdao=new GonghuoDaoImpl();
        List<Gonghuo> gonghuolist=gdao.getList();
        HttpSession session=request.getSession();
        session.setAttribute("gonghuolist",gonghuolist);
        response.sendRedirect("listgonghuo.jsp");
    }
}
```

4.4.3 购物缴费

每次在jsp页面点击"添加"或"删除"按钮都会传送参数给gouwuServlet层。执行相应的操作后再调用gouwuServlet层的jiesuan()方法,在该方法中,通过调用GouwuDaoImpl类的无参构造方法实例化GouwuDao类的对象,然后调用这个对象的jiesuan()方法,对购物表中总价列求和,将结果返回gouwuServlet层,然后gouwuServlet层将结果响应给jsp页面。

1. Model 层

(1) 定义购物类接口 GouwuDao 文件 GouwuDao.java，源代码如下：

```
public interface GouwuDao {
    public List<Gouwu> getListgouwu();
    public int addgouwu (Gouwu j);
    public int delgouwu (Gouwu j);
    public int delgouwuall();
    public float jiesuan();
}
```

(2) 定义购物操作类 GouwuDaoImpl 文件 GouwuDaoImpl.java，源代码如下：

```
package com.lx.dao;
import java.sql.Connection;
import java.sql.PreparedStatement;
import java.sql.ResultSet;
import java.sql.SQLException;
import java.util.ArrayList;
import java.util.List;
import com.lx.util.BasicJDBC;
import com.lx.entity.Gouwu;
import com.lx.entity.Kucun;
public class GouwuDaoImpl implements GouwuDao{
    private BasicJDBC db=null;
    private Connection con=null;
    public GouwuDaoImpl(){
        db=new BasicJDBC();
        con=db.getCon();
    }
    public int addgouwu(Gouwu j){
        PreparedStatement ps=null;
        String sql="insert into gouwu(name,price,type,num,sum) values(?,?,?,?,?)";
        int n=0;
        try{
            ps=con.prepareStatement(sql);
            ps.setString(1,j.getname());
            ps.setFloat(2,j.getprice());
            ps.setString(3,j.gettype());
            ps.setInt(4,j.getnum());
            ps.setFloat(5,j.getsum());
            n=ps.executeUpdate();
            System.out.println(n);
        }catch (SQLException e){
            e.printStackTrace();
```

```java
        }
        return n;
    }
    public int delgouwu(Gouwu j){
        PreparedStatement ps=null;
        String sql="delete from gouwu where name=? and price=? and num=?";
        int n=0;
        try{
            ps=con.prepareStatement(sql);
            ps.setString(1,j.getname());
            ps.setFloat(2,j.getprice());
            ps.setInt(3,j.getnum());
            n=ps.executeUpdate();
            System.out.println(n);
        }catch (SQLException e){
            e.printStackTrace();
        }
        return n;
    }
    public int delgouwuall(){
        PreparedStatement ps=null;
        String sql="delete from gouwu where name is not null";
        int n=0;
        try{
            ps=con.prepareStatement(sql);
            n=ps.executeUpdate();
            System.out.println(n);
        }catch (SQLException e){
            e.printStackTrace();
        }finally{
            db.closeAll(con,ps,null);
        }
        return n;
    }
    public float jiesuan(){
        PreparedStatement ps=null;
        String sql="select* from gouwu";
        float sum=0;
        ResultSet rs=null;
        try{
            ps=con.prepareStatement(sql);
            rs=ps.executeQuery();
            while(rs.next()){
                sum=sum+rs.getFloat("sum");
```

```java
            System.out.println(sum);
        }
    }catch(SQLException e){
        e.printStackTrace();
    }finally{
        db.closeAll(con,ps,rs);
    }
    return sum;
}
public List<Gouwu> getListgouwu(){
    PreparedStatement ps=null;
    String sql="select* from gouwu";
    ResultSet rs=null;
    List<Gouwu> list=new ArrayList<Gouwu>();
    try{
        ps=con.prepareStatement(sql);
        rs=ps.executeQuery();
        while(rs.next()){
            Gouwu j=new Gouwu(rs.getString(1),rs.getFloat(2),
                rs.getString(3),rs.getInt(4),rs.getFloat(5));
            list.add(j);
        }
    }catch(SQLException e){
        e.printStackTrace();
    }finally{
        db.closeAll(con,ps,rs);
    }
    return list;
}
}
```

2. View 层

收银页面文件，即营业员.jsp 的源代码如下：

```jsp
<%@page language="java" contentType="text/html;charset=utf-8"
    pageEncoding="utf-8"%>
<!DOCTYPE html>
<html>
<head>
<meta charset="utf-8">
<title>Insert title here</title>
</head>
<body>
<style type="text/css">
body{
```

```
        background-image:url("背景.jpg");
        background-size:100% 100%;
        background-attachment:fixed;
        }
        .text{
            width:166px;
}
div{
    position: absolute;
    top: 50%;
    left: 50%;
    -webkit-transform: translate(-50%,-50%);
    -moz-transform: translate(-50%,-50%);
    -ms-transform: translate(-50%,-50%);
    -o-transform: translate(-50%,-50%);
    transform: translate(-50%,-50%);
}
</style>
</head>
<body>
<div>
<table style="width:850px;">
<tr>
<td>
    <table style="width:150px;">
    <tr>
        <td><h2>销售营业</h2></td>
    </tr>
    <tr>
        <td>
            <ul>
                <li>
                    <a href="gouwuServlet?action=list" target="test">收银找零</a>
                </li>
            </ul>
        </td>
    </tr>
<tr>
<td><a href="login.jsp"><h2>退出登录</h2></a></td>
</tr>
</table>
</td>
<td rowspan="3" style="width:700px ;height:500px">
<iframe src="" name="test" width="100%" height="100%"></iframe>
```

```html
</tr>
</table>
</div>
</body>
</html>
```

3. Controller 层

处理购物 servlet 文件 gouwuServlet.java，源代码如下：

```java
package com.lx.servlet;
import java.io.IOException;
import java.util.List;
import javax.servlet.ServletException;
import javax.servlet.annotation.WebServlet;
import javax.servlet.http.HttpServlet;
import javax.servlet.http.HttpServletRequest;
import javax.servlet.http.HttpServletResponse;
import javax.servlet.http.HttpSession;
import com.lx.dao.GouwuDao;
import com.lx.dao.GouwuDaoImpl;
import com.lx.dao.KucunDao;
import com.lx.dao.KucunDaoImpl;
import com.lx.entity.Gouwu;
import com.lx.entity.Kucun;
import com.lx.entity.Product;
import com.lx.dao.ProductDao;
import com.lx.dao.ProductDaoImpl;
@WebServlet("/gouwuServlet")
public class gouwuServlet extends HttpServlet {
    private static final long serialVersionUID=1L;
    public gouwuServlet() {
        super();
    }
    protected void doGet(HttpServletRequest request,HttpServletResponse response)
        throws ServletException, IOException {
        doPost(request, response);
    }
    protected void doPost(HttpServletRequest request,HttpServletResponse response)
        throws ServletException, IOException {
        String action=request.getParameter("action");
        if(action.equals("list"))
        {
            listkucun(request,response);
            listgouwu(request,response);
        }
```

```java
        if(action.equals("del"))
            {
            del(request,response);
            jiesuan(request,response);
            listkucun(request,response);
            listgouwu(request,response);
            }
        if(action.equals("add"))
            {add(request,response);
            jiesuan(request,response);
            listkucun(request,response);
            listgouwu(request,response);
            }
        if(action.equals("jiesuan"))
            {
            jiesuan(request,response);
            response.sendRedirect("listgouwu.jsp");
            }
        if(action.equals("delgouwuall"))
            {
            delall(request,response);
            jiesuan(request,response);
            listgouwu(request,response);
            }
    }
    protected void add(HttpServletRequest request,HttpServletResponse response)
            throws ServletException,IOException{
            String name=request.getParameter("name");
            String price=request.getParameter("price");
            String type=request.getParameter("type");
            String num=request.getParameter("num");
            float sum=Float.parseFloat(price)* Integer.valueOf(num);
            System.out.println(name);
            System.out.println(price);
            System.out.println(num);
            Gouwu gouwu=new Gouwu(name,Float.parseFloat(price),
                type,Integer.valueOf(num),sum);
            //创建功能类对象来实现功能
            GouwuDao jdao=new GouwuDaoImpl();
            ProductDao pdao=new ProductDaoImpl();
            try{
                jdao.addgouwu(gouwu);
                pdao.updateadd(name,Integer.valueOf(num));
                pdao.updatemoneyadd(name,sum);
```

```java
            }
            catch(Exception e){
                e.printStackTrace();
                request.getRequestDispatcher("/failure.jsp").forward(request,response);
                return;
            }
        }
    protected void del(HttpServletRequest request,HttpServletResponse response)
            throws ServletException,IOException{
        String name=request.getParameter("name");
        String price=request.getParameter("price");
        String num=request.getParameter("num");
        String type=request.getParameter("type");
        String sum=request.getParameter("sum");
        System.out.println(name);
        System.out.println(price);
        System.out.println(num);
        Gouwu gouwu=new Gouwu(name,Float.parseFloat(price),type,
            Integer.valueOf(num),Float.parseFloat(sum));
            //创建功能类对象来实现功能
            GouwuDao jdao=new GouwuDaoImpl();
            ProductDao pdao=new ProductDaoImpl();
            try{
                jdao.delgouwu(gouwu);
                pdao.updatesub(name,Integer.valueOf(num));
                pdao.updatemoneysub(name,Float.parseFloat(sum));
            }
            catch(Exception e){
                e.printStackTrace();
                request.getRequestDispatcher("/failure.jsp").forward(request,response);
                return;
            }
        }
    protected void listkucun(HttpServletRequest request,HttpServletResponse response)
        throws ServletException,IOException{
        {
                KucunDao kdao=new KucunDaoImpl();
                List<Kucun> gouwulistkucun=kdao.getListkucun();
                HttpSession session=request.getSession();
                session.setAttribute("gouwulistkucun",gouwulistkucun);
        }
    }
    protected void listgouwu(HttpServletRequest request,HttpServletResponse response)
        throws ServletException,IOException{
```

```java
        {
            GouwuDao jdao=new GouwuDaoImpl();
            List<Gouwu> gouwulistgouwu=jdao.getListgouwu();
            HttpSession session=request.getSession();
            session.setAttribute("gouwulistgouwu",gouwulistgouwu);
            response.sendRedirect("listgouwu.jsp");
        }
    }
    protected void jiesuan(HttpServletRequest request,HttpServletResponse response)
        throws ServletException,IOException{
        float he=0;
        //创建功能类对象来实现功能
        GouwuDao jdao=new GouwuDaoImpl();
        try{
            he=jdao.jiesuan();
            request.getSession().setAttribute("he", he);
        }
        catch(Exception e){
            e.printStackTrace();
            request.getRequestDispatcher("/failure.jsp").forward(request,response);
            return;
        }
    }
    protected void delall(HttpServletRequest request,HttpServletResponse response)
        throws ServletException,IOException{
        //创建功能类对象来实现功能
        GouwuDao jdao=new GouwuDaoImpl();
        try{
            jdao.delgouwuall();
        }
        catch(Exception e){
            e.printStackTrace();
            request.getRequestDispatcher("/failure.jsp").forward(request,response);
            return;
        }
    }
}
```

4.4.4 产品列表

首先点击"收银找零"页面，向 gouwuServlet 层传送一个 action 参数，根据该参数的值 list，调用 gouwuServlet 层的 listkucun()方法，在该方法中通过调用 KucunDaoImpl 类的无参构造方法实例化 KucunDao 类的对象。再通过调用这个对象查询库存表中的方法，这个方法返回 gouwuServlet 层的一个数组，gouwuServlet 层再将这个数组转发给相应的 jsp 页面。该页面就会包含库存表中的数据,营业员点击对应商品后的"添加"按钮,就会将表中每个字段的值一起传送给gouwuServlet层。gouwuServlet层通过调用action参数的值选择add()方法。在add()方法中接收传送过来的参数,并定义一个变量计算单个商品的总价。将这些参数通过Gouwu类的构造方法实例化一个对象,通过GouwuDaoImpl类的无参构造方法实例化GouwuDao类的对象。调用GouwuDao对象的addgouwu()方法,就可以将数据插入购物表中,表示客户选择了这个商品。

1. Model 层

(1) 定义产品类接口 ProductDao 文件 ProductDao.java,源代码如下:

```java
package com.lx.dao;
import java.util.List;
import com.lx.entity.Product;
public interface ProductDao{
    public List <Product> getList();
    public int addProduct (Product p);
    public int delProduct (Product p);
    public int updateadd(String name,int num);
    public int updatesub(String name,int num);
    public float updatemoneyadd(String name,float money);
    public float updatemoneysub(String name,float money);
    public float jiesuan();
}
```

(2) 定义产品操作类 ProductDaoImpl 文件 ProductDaoImpl.java,源代码如下:

```java
package com.lx.dao;
import java.sql.Connection;
import java.sql.PreparedStatement;
import java.sql.ResultSet;
import java.sql.SQLException;
import java.util.ArrayList;
import java.util.List;
import com.lx.entity.Jinhuo;
import com.lx.entity.Product;
import com.lx.util.BasicJDBC;
public class ProductDaoImpl implements ProductDao{
    private BasicJDBC db=null;
    private Connection con=null;
    public ProductDaoImpl(){
        db=new BasicJDBC();
        con=db.getCon();
    }
    public int addProduct(Product p){
        PreparedStatement ps=null;
        String sql="insert into shangpin(name,price,type,amount,money) values(?,?,?,?,?)";
        int n=0;
```

```java
        try{
            ps=con.prepareStatement(sql);
            ps.setString(1,p.getname());
            ps.setFloat(2,p.getprice());
            ps.setString(3,p.gettype());
            ps.setInt(4,0);
            ps.setFloat(5,0);
            n=ps.executeUpdate();
            System.out.println(n);
        }catch (SQLException e){
            e.printStackTrace();
        }
        return n;
    }
    public int delProduct(Product p){
        PreparedStatement ps=null;
        String sql="delete from shangpin where name=? and price=? and type=?";
        int n=0;
        try{
            ps=con.prepareStatement(sql);
            ps.setString(1,p.getname());
            ps.setFloat(2,p.getprice());
            ps.setString(3,p.gettype());
            n=ps.executeUpdate();
            System.out.println(n);
        }catch (SQLException e){
            e.printStackTrace();
        }finally{
            db.closeAll(con,ps,null);
        }
        return n;
    }
    public int updateadd(String name,int num){
        PreparedStatement ps=null;
        String sql="update shangpin set amount=amount+? where name=?";
        System.out.println(name);
        int n=0;
        try{
            ps=con.prepareStatement(sql);
            ps.setInt(1,num);
            ps.setString(2,name);
            n=ps.executeUpdate();
            System.out.println(n);
        }catch (SQLException e){
            e.printStackTrace();
        }
        return n;
```

```java
    }
    public float updatemoneyadd(String name,float money){
        PreparedStatement ps=null;
        String sql="update shangpin set money=money+? where name=?";
        System.out.println(name);
        int n=0;
        try{
            ps=con.prepareStatement(sql);
            ps.setFloat(1,money);
            ps.setString(2,name);
            n=ps.executeUpdate();
            System.out.println(n);
        }catch (SQLException e){
            e.printStackTrace();
        }finally{
            db.closeAll(con,ps,null);
        }
        return n;
    }
    public int updatesub(String name,int num){
        PreparedStatement ps=null;
        String sql="update shangpin set amount=amount-? where name=?";
        System.out.println(name);
        int n=0;
        try{
            ps=con.prepareStatement(sql);
            ps.setInt(1,num);
            ps.setString(2,name);
            n=ps.executeUpdate();
            System.out.println(n);
        }catch (SQLException e){
            e.printStackTrace();
        }
        return n;
    }
    public float updatemoneysub(String name,float money){
        PreparedStatement ps=null;
        String sql="update shangpin set money=money-? where name=?";
        System.out.println(name);
        int n=0;
        try{
            ps=con.prepareStatement(sql);
            ps.setFloat(1,money);
            ps.setString(2,name);
            n=ps.executeUpdate();
            System.out.println(n);
        }catch (SQLException e){
```

```java
            e.printStackTrace();
        }finally{
            db.closeAll(con,ps,null);
        }
        return n;
    }
    public float jiesuan(){
        PreparedStatement ps=null;
        String sql="select* from shangpin";
        float sum=0;
        ResultSet rs=null;
        try{
            ps=con.prepareStatement(sql);
            rs=ps.executeQuery();
            while(rs.next()){
                sum=sum+rs.getFloat("money");
                System.out.println(sum);
            }
        }catch (SQLException e){
            e.printStackTrace();
        }finally{
            db.closeAll(con,ps,rs);
        }
        return sum;
    }
    public List<Product> getList(){
        PreparedStatement ps=null;
        String sql="select* from shangpin";
        ResultSet rs=null;
        List<Product> list=new ArrayList<Product>();
        try{
            ps=con.prepareStatement(sql);
            rs=ps.executeQuery();
            while(rs.next()){
                Product p=new Product(rs.getString(1),rs.getFloat(2),
                    rs.getString(3),rs.getInt(4),rs.getFloat(5));
                list.add(p);
            }
        }catch (SQLException e){
            e.printStackTrace();
        }finally{
            db.closeAll(con,ps,rs);
        }
        return list;
    }
}
```

2. View 层(略)

3. Controller 层

产品处理 servlet 文件 productServlet.java 的主要代码如下:

```java
package com.lx.servlet;
import java.io.IOException;
import java.util.List;
import javax.servlet.ServletException;
import javax.servlet.annotation.WebServlet;
import javax.servlet.http.HttpServlet;
import javax.servlet.http.HttpServletRequest;
import javax.servlet.http.HttpServletResponse;
import javax.servlet.http.HttpSession;
import com.lx.dao.GouwuDao;
import com.lx.dao.GouwuDaoImpl;
import com.lx.dao.ProductDao;
import com.lx.dao.ProductDaoImpl;
import com.lx.entity.Product;
@WebServlet("/productServlet")
public class productServlet extends HttpServlet{
    private static final long serialVersionUID=1L;
    public productServlet(){
        super();
    }
    protected void doGet(HttpServletRequest request,HttpServletResponse response)
        throws ServletException,IOException{
        doPost(request,response);
    }
    protected void doPost(HttpServletRequest request,HttpServletResponse response)
        throws ServletException,IOException{
        String action=request.getParameter("action");
        if(action.equals("list"))
        {
            list(request,response);
            jiesuan(request,response);
            response.sendRedirect("listxiaoshou.jsp");
        }
    }
    protected void list(HttpServletRequest request,HttpServletResponse response)
        throws ServletException,IOException{
        ProductDao pdao=new ProductDaoImpl();
        List<Product> productlist=pdao.getList();
        HttpSession session=request.getSession();
        session.setAttribute("productlist",productlist);
    }
    protected void jiesuan(HttpServletRequest request,HttpServletResponse response)
        throws ServletException,IOException{
```

```
        float xiaoshou=0;
        //创建功能类对象来实现功能
        ProductDao pdao=new ProductDaoImpl();
        try{
            xiaoshou=pdao.jiesuan();
            request.getSession().setAttribute("xiaoshou", xiaoshou);
        }
        catch(Exception e){
            e.printStackTrace();
            request.getRequestDispatcher("/failure.jsp").forward(request,response);
            return;
        }
    }
}
```

4.5 项目搭建

超市商品采购管理系统使用 eclipse 工具和 MySQL 数据库开发,系统项目结构如图 4-16、图 4-17 所示。

图 4-16 系统项目结构图 1　　　　图 4-17 系统项目结构图 2

第5章 物流快递管理系统

5.1 需求分析

5.1.1 系统概述

快递的信息管理是对物流信息的收集、整理、存储、传播和利用的过程,也就是将快递信息从分散到集中、从无序到有序、从传播到利用的过程。同时对涉及快递信息活动的各种要素,包括人员、技术、工具等进行管理,实现资源的合理配置。

物流快递管理系统最核心的需求是实现物流快递信息的收集、存储、查询、更新等功能。该系统的功能需求描述如下。

- 系统需要实现物流快递信息的在线填写及提交功能。
- 系统需要实现物流快递信息的查询功能,显示物流快递实时状态。
- 系统需要实现物流快递方面的新闻资讯信息展示功能。
- 系统需要实现物流快递涉及的业务范围信息展示功能。
- 系统需要实现在线留言功能,以方便用户向网站系统反馈信息。
- 系统需要实现管理员登录功能。
- 系统需要实现根据角色类型区分用户功能。
- 系统需要实现用户管理功能,对用户执行增、删、改、查等操作。
- 系统需要实现物流快递订单管理功能,对物流快递订单执行增、删、改、查等操作。
- 系统需要实现新闻管理功能,支持信息编辑和发布功能。
- 系统需要实现留言管理功能,支持留言编辑和删除功能。

综上可知,系统所参与的主要为用户和管理员,通过物流快递管理系统使用的不同角色及所涉及的用例,可以直观地了解两者之间的关联。在构建不同对象的功能用例分析中,更宜于理清功能全局,为下一步的设计环节打好基础。在任何一个软件系统的设计中,开发之前都要深入一线了解系统不同使用者的诉求,以满足用户对系统功能的真实需求,从而更好地设计出符合用户满意的软件产品。下面通过用户用例分析和管理员用例分析来阐述系统需求。

1. 用户

用户进入物流快递管理系统后,可以在线下单,下单完成后,可以查询订单状态;接着,用户可以浏览相关物流快递的新闻资讯,可以浏览网站提供的业务范围;最后,用户如果需要投诉或者反馈信息,可以给网站在线留言。

用户用例图如图5-1所示。

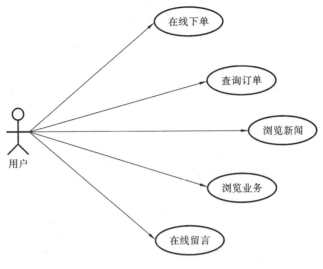

图 5-1　用户用例图

图 5-1 中各选项说明如下。

在线下单：用户可以在线下单，填写发货人、收货人和货物相关信息即可。

查询订单：用户下单后，可以输入订单号，查询物流快递订单的状态。

浏览新闻：用户可以在网站上浏览相关物流快递的最新资讯。

浏览业务：用户可以浏览网站上的业务介绍，了解业务范围，明确自己的物流快递需求。

在线留言：用户可以给网站在线留言，填写相关信息即可。

2．管理员

管理员登录系统后，可以进行菜单管理、角色管理、用户管理、订单管理、新闻管理、留言管理和查看日志。

菜单管理：管理员可以增、删、改、查菜单信息。

角色管理：对角色信息进行管理，可以增、删、改、查角色信息。

用户管理：对用户信息进行管理，可以添加、修改、查询和删除用户信息。

订单管理：对订单信息进行管理，可以添加、修改、查询和删除订单信息。

新闻管理：对新闻进行管理，可以添加、修改、查询和删除新闻资讯。

留言管理：对留言信息进行管理，可以修改和删除留言信息。

查看日志：可以查看系统的详细日志信息。

管理员用例图如图 5-2 所示。

5.1.2　功能需求描述

物流快递管理系统用例图如图 5-3 所示。

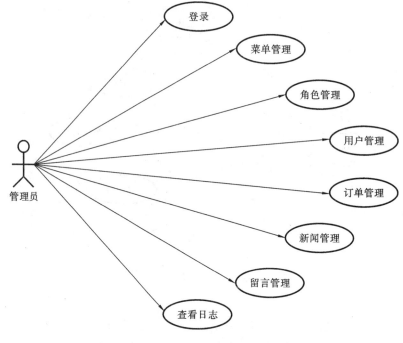

图 5-2 管理员用例图

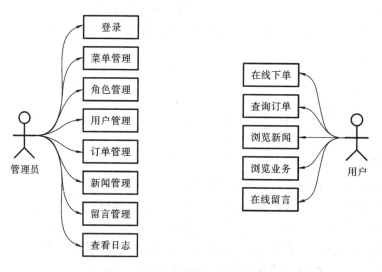

图 5-3 物流快递管理系统用例图

5.2 总体设计

5.2.1 系统总体功能结构

通过分析,将物流快递管理系统分为前台功能模块和后台功能模块。其中前台功能模块可以实现首页展示、在线下单、查询订单、业务范围、新闻资讯、在线留言和关于我们等功能。前台功能模块结构如图 5-4 所示。

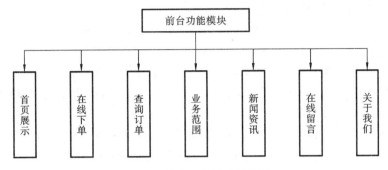

图 5-4 前台功能模块结构图

首页展示：首页显示导航栏，图片以轮播方式显示网站宣传广告。

在线下单：用户进入系统后，可以进行在线下单，以表单的形式让用户填写信息，用户需要填写发件人姓名、发件人手机、发件人地址、货物名称、货物重量、收件人姓名、收件人手机、收件人地址、备注等信息，填写完成之后，点击"立即下单"按钮即可。

查询订单：用户在线下单完成后，如果需要查询订单的物流信息，输入快递单号，就能进行查询。

业务范围：为方便用户使用物流快递服务，网站系统会详细介绍业务范围的种类，包括物流卡航、物流空运、限时到达、物流普运、城际快线等业务。

新闻资讯：为方便用户了解物流快递相关方面的信息，网站系统会显示新闻资讯，以新闻列表形式展示出来。

在线留言：为提升网站用户体验和方便用户反馈与投诉，用户可以给网站留言，用户填写姓名、联系电话、信息，然后点击"立即发送"按钮即可。

关于我们：显示物流快递公司的详细信息。

后台功能模块可以实现登录、菜单管理、角色管理、用户管理、订单管理、新闻管理、留言管理、系统日志等功能。后台功能模块结构如图 5-5 所示。

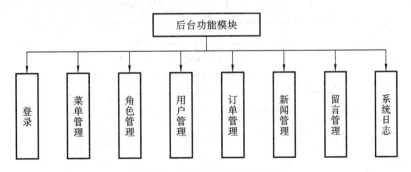

图 5-5 后台功能模块结构图

登录：管理员登录系统后台，需要输入用户名、密码、验证码，然后点击"登录"按钮即可。

菜单管理：管理员进行菜单管理，可以添加、修改、搜索和删除菜单信息。其中，添加学科信息需要填写学科名称、上级菜单、菜单 URL 和菜单图标。

角色管理：管理员进行角色管理，可以添加、修改、查询和删除角色信息。其中，添加角色信息需要填写角色名称和角色备注等。

用户管理：管理员进行用户管理，可以添加、修改、查询和删除用户信息。其中，添加用户信息需要填写用户名、密码、所属角色、性别、年龄和地址等。

订单管理：管理员进行订单管理，可以添加、修改、搜索和删除订单信息。其中，订单列表包含订单编号、发件人、发件人手机、发件人地址、收件人、收件人手机、收件人地址、货物名称、申报价格、重量、体积和时间等信息。

新闻管理：管理员进行新闻管理，可以添加、修改、搜索和删除新闻信息。其中，添加新闻信息需要填写新闻标题、所属分类、摘要、新闻标签、新闻作者、新闻内容等。

留言管理：管理员进行留言管理，可以搜索和删除留言信息。

系统日志：管理员可以查看系统所有的操作日志信息，例如登录日志。

5.2.2 总体架构

物流快递管理系统采用 SSM（Spring、Spring MVC、MyBatis）框架开发，是标准的 MVC 模式，将整个系统划分为 View 层、Controller 层、Service 层、DAO 层和持久层五层。其中，Spring MVC 负责请求的转发和视图管理，Spring 实现业务对象管理，MyBatis 作为数据对象的持久化引擎。系统架构如图 5-6 所示。

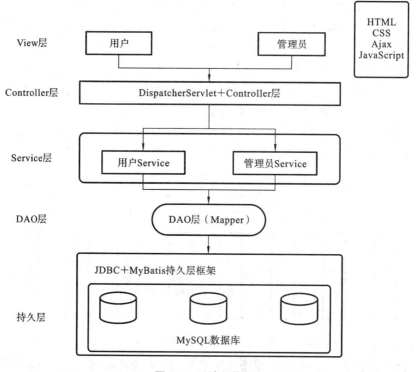

图 5-6 系统架构图

View 层:与 Controller 层结合比较紧密,需要二者结合起来协同开发,主要显示前台 jsp 页面。

Controller 层:导入 Service 层,由于用到了 Service 中的方法,所以 Controller 主要通过接收前端传送过来的参数执行业务操作,再返回指定的路径或者数据表。

Service 层:进行业务逻辑处理,执行一些关于数据库处理的操作,但不直接与数据库打交道。Service 层有接口及其实现方法,在接口的实现方法中需要导入 Dao 层。Dao 层是直接与数据库打交道的,是数据库 CRUD 的接口,只有方法名,具体实现在 mapper.xml 文件里,对数据库进行数据持久化操作。

Dao 层:负责对数据库中的数据执行增、删、改、查等操作。在它注册的框架中,如果不使用 Spring,那么每层之间的数据传递需要 new 调用该层数据类的实例。如果使用 Spring,则需要将 DAO 层和 BIZ 层的每个类都写入一个接口类,接口类里写入实现类的方法,调用时不用 new 对象,直接用对象点(.)方法就可,记住每个对象还要加上 set/get 方法。

持久层:MyBatis 是持久层框架,支持 JDBC,简化了持久层的开发。

使用 MyBatis 时,只需要通过接口指定数据操作的抽象方法,然后配置与之关联的 SQL 语句即可。

5.2.3 数据库设计

物流快递管理系统的数据库名称为 db_logistics,数据库中包括:

(1) 用户表(user):主要用于存储管理员信息。
(2) 权限表(authority):主要用于存储权限信息。
(3) 角色表(role):主要用于存储角色信息。
(4) 菜单表(menu):主要用于存储菜单信息。
(5) 订单表(orders)表:主要用于存储订单信息。
(6) 订单状态表(order_status):主要用于存储订单状态信息。
(7) 新闻资讯表(news):主要用于存储新闻资讯。
(8) 新闻资讯分类表(news_category):主要用于存储新闻资讯分类信息。
(9) 在线留言表(leave_msg):主要用于存储在线留言信息。
(10) 日志表(log):主要用于存储日志信息。

各表的数据结构如表 5-1 至表 5-10 所示。

表 5-1 用户表

字段名称	数据类型	主键	是否空	说明
id	int(11)	Y	N	管理员和用户 ID
username	varchar(32)	N	N	用户名
password	varchar(32)	N	N	密码
roleId	int(11)	N	N	角色 ID
photo	varchar(128)	N	Y	头像图片

续表

字段名称	数据类型	主键	是否空	说明
sex	int(1)	N	N	性别
age	int(3)	N	N	年龄
address	varchar(128)	N	Y	地址

表 5-2 权限表

字段名称	数据类型	主键	是否空	说明
id	int(11)	Y	N	权限 ID
roleId	int(11)	N	N	角色 ID
menuId	int(11)	N	N	菜单 ID

表 5-3 角色表

字段名称	数据类型	主键	是否空	说明
id	int(11)	Y	N	角色 ID
name	varchar(64)	N	N	角色名称
remark	varchar(128)	N	Y	备注

表 5-4 菜单表

字段名称	数据类型	主键	是否空	说明
id	int(11)	Y	N	菜单 ID
parentId	int(11)	N	N	父菜单 ID
name	varchar(32)	N	N	菜单名称
url	varchar(128)	N	Y	网址链接
icon	varchar(32)	N	N	图标

表 5-5 订单表

字段名称	数据类型	主键	是否空	说明
id	int(11)	Y	N	订单 ID
sn	varchar(32)	N	N	订单编号
sender	varchar(32)	N	N	发货人
senderMobile	varchar(32)	N	N	发货人手机
senderTel	varchar(32)	N	Y	发货人固定电话
senderAddress	varchar(128)	N	N	发货人地址
receiver	varchar(32)	N	N	收货人

续表

字段名称	数据类型	主键	是否空	说明
receiverMobile	varchar(32)	N	N	收货人手机
receiverTel	varchar(32)	N	Y	收货人固定电话
receiverAddress	varchar(128)	N	N	收货人地址
goodsName	varchar(128)	N	N	货物名称
goodsPrice	varchar(32)	N	N	货物重量
goodsWeight	varchar(32)	N	N	报价申明
goodsVolum	varchar(32)	N	N	货物体积
remark	varchar(512)	N	Y	备注信息
createTime	datetime	N	N	创建时间

表 5-6 订单状态表

字段名称	数据类型	主键	是否空	说明
id	int(11)	Y	N	订单状态 ID
orderSn	varchar(32)	N	N	订单编号
content	varchar(512)	N	Y	内容
createTime	datetime	N	N	创建时间

表 5-7 新闻资讯表

字段名称	数据类型	主键	是否空	说明
id	int(11)	Y	N	新闻资讯 ID
categoryId	int(11)	N	N	分类 ID
title	varchar(128)	N	N	标题
abstrs	varchar(512)	N	N	摘要
tags	varchar(128)	N	N	标记
photo	varchar(64)	N	N	图片
author	varchar(32)	N	N	作者
content	longtext	N	N	内容
viewNumber	int(8)	N	N	浏览数量
commentNumber	int(5)	N	N	评论数量
createTime	datetime	N	N	创建时间

表 5-8 新闻资讯分类表

字 段 名 称	数 据 类 型	主　　键	是 否 空	说　　明
id	int(11)	Y	N	新闻资讯分类 ID
name	varchar(32)	N	N	名称
sort	int(3)	N	N	序号

表 5-9 在线留言表

字 段 名 称	数 据 类 型	主　　键	是 否 空	说　　明
id	int(11)	Y	N	在线留言 ID
name	varchar(128)	N	N	姓名
tel	varchar(32)	N	N	电话
content	varchar(512)	N	N	内容
createTime	datetime	N	N	创建时间

表 5-10 日志表

字 段 名 称	数 据 类 型	主　　键	是 否 空	说　　明
id	int(11)	Y	N	日志 ID
content	varchar(255)	N	N	日志内容
createTime	datetime	N	Y	创建时间

5.3 详细设计

5.3.1 订单管理

订单管理是指管理员对订单进行管理,可以添加、修改、搜索和删除订单信息。其中,订单列表包含订单编号、发件人、发件人手机、发件人地址、收件人、收件人手机、收件人地址、货物名称、申报价格、重量、体积和时间等信息。订单管理页面如图 5-7 所示,其顺序图如图 5-8 所示。

图 5-7 订单管理页面

第 5 章 物流快递管理系统

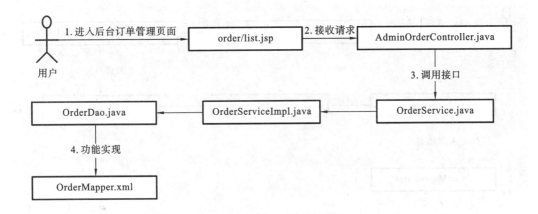

图 5-8 订单管理顺序图

进入后台订单管理页面,通过用户的增、删、改、查操作,list.jsp 页面将请求发送至 AdminOrderController,AdminOrderController 接收参数后,调用 OrderService 接口的实现类 OrderServiceImpl,即调用持久层 OrderDao 接口对数据库进行操作,然后将结果逐层返回至 list.jsp 页面并显示出来。

5.3.2 在线下单

用户进入系统后,可以进行在线下单,以表单的形式让用户填写信息,用户需要填写发货人姓名、发货人手机、发货人地址、货物名称、货物重量、收货人姓名、收货人手机、收货人地址、备注信息等,填写完成之后,点击"立即下单"按钮即可。在线下单页面如图 5-9 所示,在线下单顺序图如图 5-10 所示。

图 5-9 在线下单页面

进入后台在线下单页面,通过用户提交键入的信息,online_order.jsp 页面将请求发送

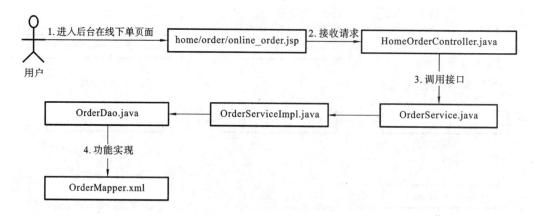

图 5-10 在线下单顺序图

至 HomeOrderController，HomeOrderController 接收参数后，调用 OrderService 接口的实现类 OrderServiceImpl，即调用持久层 OrderDao 接口对数据库进行操作，然后将结果逐层返回至 online_order.jsp 页面并显示出来。

5.3.3 查询订单

用户在线下单完成后，如果需要查询订单的物流信息，输入快递单号，就能进行查询。查询订单页面如图 5-11 所示，查询订单顺序图如图 5-12 所示。

图 5-11 查询订单页面

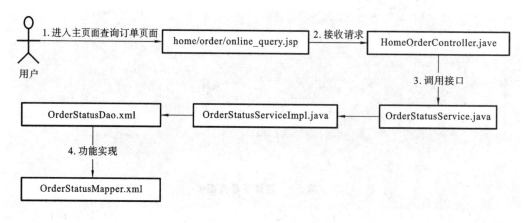

图 5-12 查询订单顺序图

进入主页面查询订单页面,通过用户的键入操作,online_query.jsp 页面将请求发送至 HomeOrderController,HomeOrderController 接收参数后,调用 OrderStatusService 接口的实现类 OrderStatusServiceImpl,即调用持久层 OrderStatusDao 接口对数据库进行操作,然后将结果逐层返回至 online_query.jsp 页面并显示出来。

5.4 编码实现

本节重点介绍公共模块、订单管理模块、在线下单模块和查询订单模块。

5.4.1 公共模块

1. 数据库连接

db.properties 文件的源代码如下:

```
datasource.connection.driver_class=com.mysql.jdbc.Driver
datasource.connection.url=jdbc:mysql://localhost:3306/db_logistics? useUnicode=
    true&characterEncoding=utf-8
datasource.connection.username=root
datasource.connection.password=1234
#连接池保持的最小连接数,default:3(建议使用)
datasource.connection.minPoolSize=3
#连接池中拥有的最大连接数,如果获得新连接时会使连接总数超过这个值,则不会再获取新连接,
而是等待其他连接释放,因此,这个值有可能会设计得很大,default:15(建议使用)
datasource.connection.maxPoolSize=15
#连接的最长空闲时间,如果超过这个时间,某个数据库连接还没有被使用,则会断开这个连接。如
果为0,则永远不会断开连接,即回收此连接。default:0(单位为 s,建议使用)
datasource.connection.maxIdleTime=0
#连接池在无空闲连接可用时,一次性创建新数据库连接数,default:3(建议使用)
datasource.connection.acquireIncrement=3
#连接池为数据源缓存的 PreparedStatement 总数。由于 PreparedStatement 属于单个 Connec-
tion,所以这个数量应该根据应用中平均连接数乘以每个连接的平均 PreparedStatement 来计算。
同时,maxStatementsPerConnection 的配置无效。default:0(不建议使用)
datasource.connection.maxStatements=0
#连接池为数据源单个 Connection 缓存的 PreparedStatement 数,该配置比 maxStatements 的更
有意义,因为它缓存的服务对象是单个数据连接,如果设置得好,那么是可以提高性能的。为0时不
缓存。default:0(看情况而论)
datasource.connection.maxStatementsPerConnection=0
#连接池初始化时创建的连接数,default:3(建议使用)
datasource.connection.initialPoolSize=3
#用来配置测试空闲连接的间隔时间。测试方式使用上面两种之一,可以用来解决 MySQL 8 小时断
开连接的问题。因为它保证连接池会每隔一定时间对空闲连接进行一次测试,从而保证有效的空闲
连接能每隔一定时间访问一次数据库,将 MySQL 8 小时无会话的状态打破。为0时不测试。
default:0(建议使用)
datasource.connection.idleConnectionTestPeriod=0
```

#连接池在获得新连接失败时重试的次数,如果小于等于0,则无限重试直至连接获得成功。
default:30(建议使用)
datasource.connection.acquireRetryAttempts=30
#如果为true,当连接获取失败时,则自动关闭数据源,除非重新启动应用程序,所以一般不用。
default:false(不建议使用)
datasource.connection.breakAfterAcquireFailure=false
#性能消耗大。如果为true,当每次getConnection时都会测试,为了提高性能,尽量不要使用。
default:false(不建议使用)
datasource.connection.testConnectionOnCheckout=false
#配置当连接池的所有连接用完时,应用程序getConnection的等待时间。如果为0,则无限等待直至有其他连接释放或者创建新的连接,如果不为0,当时间到的时候仍没有获得连接,就会抛出SQLException。也就是acquireRetryAttempts*acquireRetryDelay。default:0(与上面两个有重复,选择其中两个都可)
datasource.connection.checkoutTimeout=30000
#如果为true,则在关闭的时候测试连接的有效性。default:false(不建议使用)
datasource.connection.testConnectionOnCheckin=false
#配置一个表名,连接池根据这个表名使用自己的SQL语句在这个空表上测试数据库连接,这个表只能由c3p0来使用,用户不能操作。default:null(不建议使用)
datasource.connection.automaticTestTable=c3p0TestTable
#连接池在获得新连接时的间隔时间。default:1000的单位为ms(建议使用)
datasource.connection.acquireRetryDelay=1000
#如果为0,则所有的Connection在应用程序中必须关闭。如果不为0,则强制在设定的时间到达后回收Connection,所以必须小心设置,保证在回收之前所有数据库操作都能够完成。这种限制应减少,Connection未关闭的情况下不太适用。建议手动关闭。default:0的单位为s(不建议使用)
datasource.connection.unreturnedConnectionTimeout=0
#这个配置主要是为了快速减轻连接池的负载,比如在连接池中,连接数因为某次数据访问高峰导致创建了很多数据库连接,但是后面的时间段需要的数据库连接数很少,需要快速释放,所以必须小于maxIdleTime。实际上没必要配置,因为maxIdleTime已经配置了。default:0的单位为s(不建议使用)
datasource.connection.maxIdleTimeExcessConnections=0
#配置连接的生存时间,超过这个时间的连接将由连接池自动断开丢弃。当然,正在使用的连接不会马上断开,而是等待它关闭再断开。如果配置为0,则不会对连接的生存时间进行限制。
default:0的单位为s(不建议使用)
datasource.connection.maxConnectionAge=0

2. 拦截器

LoginInterceptor.java文件的源代码如下:

```
package com.ischoolbar.programmer.interceptor.admin;
import java.util.HashMap;
import java.util.List;
import java.util.Map;
import javax.servlet.http.HttpServletRequest;
import javax.servlet.http.HttpServletResponse;
import net.sf.json.JSONObject;
```

```java
import org.apache.commons.lang.StringUtils;
import org.springframework.web.servlet.HandlerInterceptor;
import org.springframework.web.servlet.ModelAndView;
import com.ischoolbar.programmer.entity.admin.Menu;
import com.ischoolbar.programmer.util.MenuUtil;
/**
 * 后台登录拦截器
 */
public class LoginInterceptor implements HandlerInterceptor {
    public void afterCompletion(HttpServletRequest arg0,
            HttpServletResponse arg1, Object arg2, Exception arg3)
            throws Exception {
    }
    public void postHandle(HttpServletRequest arg0, HttpServletResponse arg1,
            Object arg2, ModelAndView arg3) throws Exception {
    }
    public boolean preHandle(HttpServletRequest request, HttpServletResponse response,
            Object arg2) throws Exception {
        String requestURI = request.getRequestURI();
        Object admin = request.getSession().getAttribute("admin");
        if(admin == null){
            //表示未登录或者登录失效
            System.out.println("连接"+requestURI+"进入拦截器!");
            String header = request.getHeader("X-Requested-With");
            //判断是否是ajax请求
            if("XMLHttpRequest".equals(header)){
                //表示是ajax请求
                Map<String, String> ret = new HashMap<String, String>();
                ret.put("type", "error");
                ret.put("msg", "登录会话超时或还未登录,请重新登录!");
                response.getWriter().write(JSONObject.fromObject(ret).toString());
                return false;
            }
            //表示是普通连接跳转,直接重定向到登录页面
            response.sendRedirect(request.getServletContext().getContextPath()+
                "/system/login");
            return false;
        }
        //获取菜单id
        String mid = request.getParameter("_mid");
        if(!StringUtils.isEmpty(mid)){
            List<Menu> allThirdMenu = MenuUtil.getAllThirdMenu((List<Menu>)
                request.getSession().getAttribute("userMenus"), Long.valueOf(mid));
            request.setAttribute("thirdMenuList", allThirdMenu);
```

```
            }
            return true;
        }
}
```

3. 配置文件

（1）配置文件 web.xml 的源代码如下：

```xml
<?xml version="1.0" encoding="UTF-8"?>
<web-app xmlns:xsi="http://www.w3.org/2001/XMLSchema-instance" xmlns=
    "http://java.sun.com/xml/ns/javaee" xmlns:web=
    "http://java.sun.com/xml/ns/javaee/web-app_2_5.xsd"
    xsi:schemaLocation="http://java.sun.com/xml/ns/javaee
    http://java.sun.com/xml/ns/javaee/web-app_2_5.xsd" id=
    "WebApp_ID" version="2.5">
    <!-- 中文乱码处理 -->
    <filter>
        <filter-name>CharacterEncodingFilter</filter-name>
        <filter-class>org.springframework.web.filter.CharacterEncodingFilter
        </filter-class>
        <init-param>
            <param-name>encoding</param-name>
        <param-value>UTF-8</param-value>
        </init-param>
        <init-param>
            <param-name>forceEncoding</param-name>
            <param-value>true</param-value>
        </init-param>
    </filter>
    <filter-mapping>
        <filter-name>CharacterEncodingFilter</filter-name>
        <url-pattern>/*</url-pattern>
    </filter-mapping>
    <filter>
        <filter-name>HiddenHttpMethodFilter</filter-name>
   <filter-class>org.springframework.web.filter.HiddenHttpMethodFilter</filter-class>
        </filter>
    <filter-mapping>
        <filter-name>HiddenHttpMethodFilter</filter-name>
        <url-pattern>/*</url-pattern>
    </filter-mapping>
    <!-- Spring 配置文件信息 -->
    <context-param>
        <param-name>contextConfigLocation</param-name>
```

```xml
            <param-value>classpath:config/spring/applicationContext.xml</param-value>
    </context-param>
    <!-- ContextLoaderListener 监听器 -->
    <listener>
        <listener-class>org.springframework.web.context.ContextLoaderListener
        </listener-class>
    </listener>
    <!-- 日志配置 -->
    <context-param>
        <param-name>log4jConfigLocation</param-name>
        <param-value>classpath:config/log4j.properties</param-value>
    </context-param>
    <listener>
        <listener-class>org.springframework.web.util.Log4jConfigListener
        </listener-class>
    </listener>
    <!-- 配置前端控制器 -->
    <servlet>
        <servlet-name>DispatcherServlet</servlet-name>
        <servlet-class>org.springframework.web.servlet.DispatcherServlet
        </servlet-class>
        <init-param>
            <param-name>contextConfigLocation</param-name>
            <param-value>classpath:config/springmvc/springmvc.xml</param-value>
        </init-param>
        <load-on-startup>1</load-on-startup>
    </servlet>
    <servlet-mapping>
        <servlet-name>DispatcherServlet</servlet-name>
        <url-pattern>/</url-pattern>
    </servlet-mapping>
    <error-page>
        <error-code>404</error-code>
        <location>/WEB-INF/errors/404.jsp</location>
    </error-page>
    <error-page>
        <error-code>500</error-code>
        <location>/WEB-INF/errors/500.jsp</location>
    </error-page>
      <welcome-file-list>
        <welcome-file>index.jsp</welcome-file>
      </welcome-file-list>
  <context-param>
     <param-name>webAppRootKey</param-name>
```

```xml
        <param-value>map.root</param-value>
    </context-param>
</web-app>
```

（2）applicationContext.xml文件的源代码如下：

```xml
<?xml version="1.0" encoding="UTF-8"?>
<beans xmlns="http://www.springframework.org/schema/beans"
    xmlns:xsi="http://www.w3.org/2001/XMLSchema-instance" xmlns:context=
        "http://www.springframework.org/schema/context"
    xmlns:tx="http://www.springframework.org/schema/tx"
    xsi:schemaLocation="http://www.springframework.org/schema/beans
        http://www.springframework.org/schema/beans/spring-beans-4.2.xsd
        http://www.springframework.org/schema/context
        http://www.springframework.org/schema/context/spring-context-4.2.xsd
        http://www.springframework.org/schema/tx
        http://www.springframework.org/schema/tx/spring-tx-4.2.xsd">
    <context:component-scan base-package="com.ischoolbar.programmer">
        <context:include-filter type="annotation"
            expression="org.springframework.stereotype.Component" />
        <context:include-filter type="annotation"
            expression="org.springframework.stereotype.Repository" />
        <context:include-filter type="annotation"
            expression="org.springframework.stereotype.Service" />
    </context:component-scan>
    <!--加载数据源配置文件 db.properties -->
    <context:property-placeholder location="classpath:config/db.properties" />
    <!--配置c3p0数据源 -->
    <bean id="dataSource" class="com.mchange.v2.c3p0.ComboPooledDataSource"
        destroy-method="close">
        <property name="driverClass" value="${datasource.connection.driver_class}"/>
        <property name="jdbcUrl" value="${datasource.connection.url}"/>
        <property name="user" value="${datasource.connection.username}"/>
        <property name="password" value="${datasource.connection.password}"/>
        <property name="minPoolSize" value="${datasource.connection.minPoolSize}"/>
        <!-- 连接池中保留的最大连接数。Default:15 -->
        <property name="maxPoolSize" value=
            "${datasource.connection.maxPoolSize}"/>
        <!--最大空闲时间,如果60秒内未使用,则连接被丢弃。若为0,则永不丢弃。
            Default:0-->
        <property name="maxIdleTime" value=
            "${datasource.connection.maxIdleTime}"/>
        <!-- 当连接池中的连接耗尽时c3p0一次同时获取的连接数。Default:3 -->
        <property name="acquireIncrement" value=
            "${datasource.connection.acquireIncrement}"/>
```

```xml
<!-- JDBC 的标准参数,用于控制数据源内加载的 PreparedStatements 数量。
但由于预缓存的 statements 属于单个 connection 而不是整个连接池,所以设置这个参数需要考
虑多方面的因素。
如果 maxStatements 与 maxStatementsPerConnection 均为 0,则缓存被关闭。
Default:0-->
<property name="maxStatements" value=
    "${datasource.connection.maxStatements}"/>
<!-- maxStatementsPerConnection 定义了连接池内单个连接所拥有的最大缓存 statements
    数。Default:0-->
<property name="maxStatementsPerConnection"
    value="${datasource.connection.maxStatementsPerConnection}"/>
<!-- 初始化时获取三个连接,取值应在 minPoolSize 与 maxPoolSize 之间。Default:3-->
<property name="initialPoolSize" value=
    "${datasource.connection.initialPoolSize}"/>
<!-- 每 60 秒检查所有连接池中的空闲连接。Default:0-->
<property name="idleConnectionTestPeriod"
    value="${datasource.connection.idleConnectionTestPeriod}"/>
<!-- 定义从数据库获取新连接失败后重复尝试的次数。Default:30-->
<property name="acquireRetryAttempts"
    value="${datasource.connection.acquireRetryAttempts}"/>
<!-- 获取连接失败将会引起所有等待连接池来获取连接的线程抛出异常。但是数据源仍
有效保留,并在下次调用 getConnection()的时候继续尝试获取连接。如果设为 true,那么在尝试
获取连接失败后,该数据源将申明已断开并永久关闭。Default: false -->
<property name="breakAfterAcquireFailure"
    value="${datasource.connection.breakAfterAcquireFailure}"/>
<!-- 因为性能消耗大,所以请只在需要的时候使用它。如果设为 true,那么在每个 con-
nection 提交的时候将校验其有效性。建议使用 idleConnectionTestPeriod 或 automaticTest-
Table 等方法来提升连接测试的性能。Default: false -->
<property name="testConnectionOnCheckout"
    value="${datasource.connection.testConnectionOnCheckout}"/>
<property name="checkoutTimeout" value=
    "${datasource.connection.checkoutTimeout}"/>
<property name="testConnectionOnCheckin"
    value="${datasource.connection.testConnectionOnCheckin}"/>
<property name="automaticTestTable" value=
    "${datasource.connection.automaticTestTable}"/>
<property name="acquireRetryDelay" value=
    "${datasource.connection.acquireRetryDelay}"/>
<!-- 自动超时回收 Connection -->
<property name="unreturnedConnectionTimeout" value=
    "${datasource.connection.unreturnedConnectionTimeout}"/>
<!-- 超时自动断开 -->
<property name="maxIdleTimeExcessConnections" value=
    "${datasource.connection.maxIdleTimeExcessConnections}"/>
```

```xml
        <property name="maxConnectionAge" value=
            "${datasource.connection.maxConnectionAge}"/>
    </bean>
    <!-- 事务管理器(JDBC) -->
    <bean id="transactionManager"
        class="org.springframework.jdbc.datasource.DataSourceTransactionManager">
        <property name="dataSource" ref="dataSource"></property>
    </bean>
    <!-- 启动声明式事务驱动 -->
    <tx:annotation-driven transaction-manager="transactionManager"/>
    <!-- spring通过sqlSessionFactoryBean获取sqlSessionFactory工厂类 -->
    <bean id="sqlSessionFactory" class="org.mybatis.spring.SqlSessionFactoryBean">
        <property name="dataSource" ref="dataSource"></property>
        <!-- 扫描po包,使用别名 -->
        <property name="typeAliasesPackage" value=
            "com.ischoolbar.programmer.entity"></property>
        <!-- 扫描映射文件 -->
        <property name="mapperLocations" value=
            "classpath:config/mybatis/mapper/admin/*.xml"></property>
    </bean>
    <!--配置扫描dao包,动态实现dao接口,注入spring容器中 -->
    <bean class="org.mybatis.spring.mapper.MapperScannerConfigurer">
        <property name="basePackage" value="com.ischoolbar.programmer.dao" />
        <!-- 注意,使用sqlSessionFactoryBeanName避免出现spring扫描组件失效问题 -->
        <property name="sqlSessionFactoryBeanName" value="sqlSessionFactory" />
    </bean>
    <bean id="gson" class="com.google.gson.Gson" scope="prototype"></bean>
</beans>
```

(3) 配置文件spingmvc.xml文件的源代码如下:

```xml
<!--只需要扫描包中的Controller注解 -->
<context:component-scan base-package="com.ischoolbar.programmer.controller">
    <context:include-filter type="annotation"
        expression="org.springframework.stereotype.Controller" />
</context:component-scan>
<!--启动mvc注解驱动 -->
<mvc:annotation-driven></mvc:annotation-driven>
<!--启动定时任务 -->
<task:annotation-driven/>
<!--静态资源处理 -->
<mvc:default-servlet-handler/>
<!--配置视图解析器 -->
<bean class="org.springframework.web.servlet.view.
    InternalResourceViewResolver">
```

```xml
        <property name="prefix" value="/WEB-INF/views/"></property>
        <property name="suffix" value=".jsp"></property>
    </bean>
    <!--文件上传 -->
    <bean id="multipartResolver"
    class="org.springframework.web.multipart.commons.CommonsMultipartResolver">
        <!-- 上传文件大小限制 -->
        <property name="maxUploadSize">
            <value>10485760</value>
        </property>
        <!--请求的编码格式,与jsp页面一致 -->
        <property name="defaultEncoding">
            <value>UTF-8</value>
        </property>
    </bean>
    <!--后台访问拦截器 -->
    <!-- -->
    <mvc:interceptors>
    <mvc:interceptor>
        <mvc:mapping path="/admin/**"/>
            <mvc:mapping path="/system/*"/>
            <mvc:exclude-mapping path="/system/login"/>
            <mvc:exclude-mapping path="/system/get_cpacha"/>
            <mvc:exclude-mapping path="/resources/**"/>
            <bean class="com.ischoolbar.programmer.interceptor.admin.
            LoginInterceptor"></bean>
        </mvc:interceptor>
    </mvc:interceptors>
</beans>
```

（4）log4j.properties 文件的源代码如下：

```
###direct log message to stdout ###
log4j.appender.stdout.Target=System.out
log4j.appender.stdout=org.apache.log4j.ConsoleAppender
log4j.appender.stdout.layout=org.apache.log4j.PatternLayout
log4j.appender.stdout.layout.ConversionPattern=
    %d{ABSOLUTE} %5p %c{1}:%L - %m%n
log4j.rootLogger=INFO, stdout
#debug,info,warning,error
```

（5）mybatis-config.xml 文件的源代码如下：

```
<?xml version="1.0" encoding="UTF-8"?>
<!DOCTYPE configuration
    PUBLIC "-//mybatis.org//DTD Config 3.0//EN"
```

```
                "http://mybatis.org/dtd/mybatis-3-config.dtd">
<configuration>
    <!-- 暂时不需要做任何配置 -->
</configuration>
```

5.4.2 订单管理

1. 持久化层

(1) Order.java 文件的源代码如下：

```
package com.ischoolbar.programmer.entity.admin;
import java.util.Date;
import org.springframework.stereotype.Component;
/**
 * 订单实体
 */
@Component
public class Order {
    private Long id;
    private String sn;                  //订单编号
    private String sender;              //发货人
    private String senderMobile;        //发货人手机
    private String senderTel;           //发货人固定电话
    private String senderAddress;       //发货人地址
    private String reciever;            //收件人
    private String recieverMobile;      //收货人手机
    private String recieverTel;         //收货人固定电话
    private String recieverAddress;     //收货人地址
    private String goodsName;           //货物名称
    private String goodsPrice;          //申报价格
    private String goodsWeight;         //重量
    private String goodsVolum;          //体积
    private String remark;              //备注信息
    private Date createTime;            //时间
    //省略 getter、setter 方法
}
```

(2) OrderStatus.java 文件的源代码如下：

```
package com.ischoolbar.programmer.entity.admin;
import java.util.Date;
import org.springframework.stereotype.Component;
/**
 * 订单追踪状态
 */
```

```java
@Component
public class OrderStatus {
    private Long id;
    private String orderSn;         //订单 id
    private String content;         //内容
    private Date createTime;        //创建时间
//省略 getter、setter 方法
}
```

2. DAO 层

(1) OrderDao.java 文件的源代码如下:

```java
package com.ischoolbar.programmer.dao.admin;
import java.util.List;
import java.util.Map;
import org.springframework.stereotype.Repository;
import com.ischoolbar.programmer.entity.admin.Order;
/**
 * 订单类 dao
 */
@Repository
public interface OrderDao {
    public int add(Order order);
    public int edit(Order order);
    public List<Order> findList(Map<String, Object> queryMap);
    public int getTotal(Map<String, Object> queryMap);
    public Order findBySn(String sn);
}
```

add()方法用于在后台添加订单；edit(Order order)方法用于编辑订单；findList()方法用于完成查询订单列表；getTotal()方法用于返回订单数；findBySn()方法用于查询某个订单。

(2) OrderMapper.xml 文件的源代码如下:

```xml
<?xml version="1.0" encoding="UTF-8"?>
<!DOCTYPE mapper
PUBLIC "-//mybatis.org//DTD Mapper 3.0//EN"
"http://mybatis.org/dtd/mybatis-3-mapper.dtd">
<mapper namespace="com.ischoolbar.programmer.dao.admin.OrderDao">
    <!-- 订单插入操作 -->
    <insert id="add" useGeneratedKeys="true" keyProperty="id" parameterType=
        "com.ischoolbar.programmer.entity.admin.Order">
    insert into orders(id,sn,sender,senderMobile,senderTel,senderAddress,reciever,
        recieverMobile,recieverTel,recieverAddress,goodsName,goodsPrice,goodsWeight,
        goodsVolum,remark,createTime)
```

```xml
        values(null,#{sn},#{sender},#{senderMobile},#{senderTel},#{senderAddress},
    #{reciever},#{recieverMobile},#{recieverTel},#{recieverAddress},#{goodsName},
    #{goodsPrice},#{goodsWeight},#{goodsVolum},#{remark},#{createTime})
</insert>
<!-- 订单更新操作 -->
<update id="edit" parameterType="com.ischoolbar.programmer.entity.admin.Order">
    update orders set sender=#{sender},senderMobile=#{senderMobile},
    senderTel=#{senderTel},senderAddress=#{senderAddress},reciever=
    #{reciever},recieverMobile=#{recieverMobile},recieverTel=#{recieverTel},
    recieverAddress=#{recieverAddress},goodsName=#{goodsName},
    goodsPrice=#{goodsPrice},goodsWeight=#{goodsWeight},goodsVolum=
    #{goodsVolum},remark=#{remark} where id=#{id}
</update>
<!-- 订单信息搜索查询 -->
<select id="findBySn" parameterType="Map" resultType=
    "com.ischoolbar.programmer.entity.admin.Order">
    select * from orders where sn='${value}'
</select>
<!-- 订单信息搜索查询 -->
<select id="findList" parameterType="Map" resultType=
    "com.ischoolbar.programmer.entity.admin.Order">
    select * from orders where 1=1
    <if test="sender != null">
        and sender like '%${sender}%'
    </if>
    <if test="senderMobile != null">
        and senderMobile like '%${senderMobile}%'
    </if>
    <if test="senderTel != null">
        and senderTel like '%${senderTel}%'
    </if>
    <if test="senderAddress != null">
        and senderAddress like '%${senderAddress}%'
    </if>
    <if test="reciever != null">
        and reciever like '%${reciever}%'
    </if>
    <if test="recieverMobile != null">
        and recieverMobile like '%${recieverMobile}%'
    </if>
    <if test="recieverTel != null">
        and recieverTel like '%${recieverTel}%'
    </if>
    <if test="recieverAddress != null">
```

```xml
            and recieverAddress like '%${recieverAddress}%'
        </if>
        <if test="goodsName!=null">
            and goodsName like '%${goodsName}%'
        </if>
        order by createTime desc
        <if test="offset!=null and pageSize!=null">
            limit #{offset},#{pageSize}
        </if>
    </select>
    <!-- 模糊搜索总条数 -->
    <select id="getTotal" parameterType="Map" resultType="Integer">
        select count(*) from orders where 1=1
        <if test="sender!=null">
            and sender like '%${sender}%'
        </if>
        <if test="senderMobile!=null">
            and senderMobile like '%${senderMobile}%'
        </if>
        <if test="senderTel!=null">
            and senderTel like '%${senderTel}%'
        </if>
        <if test="senderAddress!=null">
            and senderAddress like '%${senderAddress}%'
        </if>
        <if test="reciever!=null">
            and reciever like '%${reciever}%'
        </if>
        <if test="recieverMobile!=null">
            and recieverMobile like '%${recieverMobile}%'
        </if>
        <if test="recieverTel!=null">
            and recieverTel like '%${recieverTel}%'
        </if>
        <if test="recieverAddress!=null">
            and recieverAddress like '%${recieverAddress}%'
        </if>
        <if test="goodsName!=null">
            and goodsName like '%${goodsName}%'
        </if>
    </select>
    <!-- 删除订单信息 -->
    <delete id="delete" parameterType="String">
        delete from orders where id in(${value})
```

```
        </delete>
</mapper>
```

(3) OrderStatusDao.java 文件的源代码如下:

```
package com.ischoolbar.programmer.dao.admin;
import java.util.List;
import java.util.Map;
import org.springframework.stereotype.Repository;
import com.ischoolbar.programmer.entity.admin.OrderStatus;
/**
 * 订单状态类 dao
 */
@Repository
public interface OrderStatusDao {
    public int add(OrderStatus orderStatus);
    public List<OrderStatus> findList(Map<String, Object> queryMap);
    public int getTotal(Map<String, Object> queryMap);
}
```

(4) OrderStatusMapper.xml 文件的源代码如下:

```
<?xml version="1.0" encoding="UTF-8"?>
<!DOCTYPE mapper
PUBLIC "-//mybatis.org//DTD Mapper 3.0//EN"
"http://mybatis.org/dtd/mybatis-3-mapper.dtd">
<mapper namespace="com.ischoolbar.programmer.dao.admin.OrderStatusDao">
    <!-- 订单状态插入操作 -->
    <insert id="add" useGeneratedKeys="true" keyProperty="id" parameterType=
        "com.ischoolbar.programmer.entity.admin.OrderStatus">
        insert into order_status(id,orderSn,content,createTime) values(
            null,#{orderSn},#{content},#{createTime})
    </insert>
    <!-- 订单信息搜索查询 -->
    <select id="findList" parameterType="Map" resultType=
        "com.ischoolbar.programmer.entity.admin.OrderStatus">
        select * from order_status
        <if test="orderSn != null">
            where orderSn=#{orderSn}
        </if>
        order by createTime desc
        <if test="offset != null and pageSize != null">
            limit #{offset},#{pageSize}
        </if>
    </select>
    <!-- 模糊搜索总条数 -->
```

```xml
<select id="getTotal" parameterType="Map" resultType="Integer">
    select count(*) from order_status
    <if test="orderSn !=null">
        where orderSn=#{orderSn}
    </if>
</select>
<!-- 删除订单信息 -->
<delete id="delete" parameterType="String">
    delete from order_status where id in(${value})
</delete>
</mapper>
```

3. Service 层

（1）OrderService.java 文件的源代码如下：

```java
package com.ischoolbar.programmer.service.admin;
import java.util.List;
import java.util.Map;
import org.springframework.stereotype.Service;
import com.ischoolbar.programmer.entity.admin.Order;
/**
 * 订单接口
 */
@Service
public interface OrderService {
    public int add(Order order);
    public int edit(Order order);
    public List<Order> findList(Map<String, Object> queryMap);
    public int getTotal(Map<String, Object> queryMap);
    public Order findBySn(String sn);
}
```

（2）OrderServiceImpl.java 文件的源代码如下：

```java
package com.ischoolbar.programmer.service.admin.impl;
import java.util.List;
import java.util.Map;
import org.springframework.beans.factory.annotation.Autowired;
import org.springframework.stereotype.Service;
import com.ischoolbar.programmer.dao.admin.OrderDao;
import com.ischoolbar.programmer.entity.admin.Order;
import com.ischoolbar.programmer.service.admin.OrderService;
@Service
public class OrderServiceImpl implements OrderService {
    @Autowired
    private OrderDao orderDao;
```

```java
    @Override
    public int add(Order order) {
        return orderDao.add(order);
    }
    @Override
    public int edit(Order order) {
        return orderDao.edit(order);
    }
    @Override
    public List<Order> findList(Map<String, Object> queryMap) {
        return orderDao.findList(queryMap);
    }
    @Override
    public int getTotal(Map<String, Object> queryMap) {
        return orderDao.getTotal(queryMap);
    }
    @Override
    public Order findBySn(String sn) {
        return orderDao.findBySn(sn);
    }
}
```

(3) OrderStatusService.java 文件的源代码如下：

```java
package com.ischoolbar.programmer.service.admin;
import java.util.List;
import java.util.Map;
import org.springframework.stereotype.Service;
import com.ischoolbar.programmer.entity.admin.OrderStatus;
/**
 * 订单状态接口
 */
@Service
public interface OrderStatusService {
    public int add(OrderStatus orderStatus);
    public List<OrderStatus> findList(Map<String, Object> queryMap);
    public int getTotal(Map<String, Object> queryMap);
}
```

(4) OrderStatusServiceImpl.java 文件的源代码如下：

```java
package com.ischoolbar.programmer.service.admin.impl;
import java.util.List;
import java.util.Map;
import org.springframework.beans.factory.annotation.Autowired;
import org.springframework.stereotype.Service;
```

```java
import com.ischoolbar.programmer.dao.admin.OrderStatusDao;
import com.ischoolbar.programmer.entity.admin.OrderStatus;
import com.ischoolbar.programmer.service.admin.OrderStatusService;
@Service
public class OrderStatusServiceImpl implements OrderStatusService {
    @Autowired
    private OrderStatusDao orderStatusDao;
    @Override
    public int add(OrderStatus orderStatus) {
        return orderStatusDao.add(orderStatus);
    }
    @Override
    public List<OrderStatus> findList(Map<String, Object> queryMap) {
        return orderStatusDao.findList(queryMap);
    }
    @Override
    public int getTotal(Map<String, Object> queryMap) {
        return orderStatusDao.getTotal(queryMap);
    }
}
```

4. Controller 层

AdminOrderController.java 文件的源代码如下：

```java
package com.ischoolbar.programmer.controller.admin;
import java.util.Date;
import java.util.HashMap;
import java.util.Map;
import org.springframework.beans.factory.annotation.Autowired;
import org.springframework.stereotype.Controller;
import org.springframework.web.bind.annotation.RequestMapping;
import org.springframework.web.bind.annotation.RequestMethod;
import org.springframework.web.bind.annotation.RequestParam;
import org.springframework.web.bind.annotation.ResponseBody;
import org.springframework.web.servlet.ModelAndView;
import com.ischoolbar.programmer.entity.admin.Order;
import com.ischoolbar.programmer.entity.admin.OrderStatus;
import com.ischoolbar.programmer.page.admin.Page;
import com.ischoolbar.programmer.service.admin.OrderService;
import com.ischoolbar.programmer.service.admin.OrderStatusService;
/**
 * 订单管理控制器
 */
@RequestMapping("/admin/order")
```

```java
@Controller
public class AdminOrderController {
    @Autowired
    private OrderService orderService;
    @Autowired
    private OrderStatusService orderStatusService;
    /**
     *订单列表页面
     *@param model
     *@return
     */
    @RequestMapping(value="/list",method=RequestMethod.GET)
    public ModelAndView list(ModelAndView model){
        model.setViewName("order/list");
        return model;
    }
    /**
     *获取订单列表
     *@param page
     *@param content
     *@param roleId
     *@param sex
     *@return
     */
    @RequestMapping(value="/list",method=RequestMethod.POST)
    @ResponseBody
    public Map<String,Object> getList(Page page,
        @RequestParam(name="sender",required=false,defaultValue="") String sender,
        @RequestParam(name="senderMobile",required=false,defaultValue="")
            String senderMobile,
        @RequestParam(name="reciever",required=false,defaultValue="") String reciever,
        @RequestParam(name="recieverMobile",required=false,defaultValue="")
            String recieverMobile
        ){
        Map<String,Object> ret=new HashMap<String,Object>();
        Map<String,Object> queryMap=new HashMap<String,Object>();
        queryMap.put("sender", sender);
        queryMap.put("senderMobile", senderMobile);
        queryMap.put("reciever", reciever);
        queryMap.put("recieverMobile", recieverMobile);
        queryMap.put("offset", page.getOffset());
        queryMap.put("pageSize", page.getRows());
        ret.put("rows", orderService.findList(queryMap));
        ret.put("total", orderService.getTotal(queryMap));
```

```java
        return ret;
}
/**
 * 获取订单状态列表
 * @param page
 * @param orderSn
 * @return
 */
@RequestMapping(value="/status_list",method=RequestMethod.POST)
@ResponseBody
public Map<String, Object> getStatusList(Page page,
    @RequestParam(name="orderSn",required=false,defaultValue="")
        String orderSn
    ){
    Map<String, Object> ret=new HashMap<String, Object>();
    Map<String, Object> queryMap=new HashMap<String, Object>();
    queryMap.put("orderSn", orderSn);
    queryMap.put("offset", page.getOffset());
    queryMap.put("pageSize", page.getRows());
    ret.put("rows", orderStatusService.findList(queryMap));
    ret.put("total", orderStatusService.getTotal(queryMap));
    return ret;
}
/**
 * 添加订单
 * @param user
 * @return
 */
@RequestMapping(value="/add",method=RequestMethod.POST)
@ResponseBody
public Map<String, String> add(Order order){
    Map<String, String> ret=new HashMap<String, String>();
    if(order==null){
        ret.put("type", "error");
        ret.put("msg", "请填写正确的订单信息!");
        return ret;
    }
    order.setSn(System.currentTimeMillis()+"");
    order.setCreateTime(new Date());
    if(orderService.add(order)<=0){
        ret.put("type", "error");
        ret.put("msg", "订单添加失败,请联系管理员!");
        return ret;
    }
```

```java
            ret.put("type", "success");
            ret.put("msg", "订单添加成功!");
            return ret;
    }
    /**
     *编辑订单
     * @param order
     * @return
     */
    @RequestMapping(value="/edit",method=RequestMethod.POST)
    @ResponseBody
    public Map<String, String> edit(Order order){
        Map<String, String> ret=new HashMap<String, String>();
        if(order==null){
            ret.put("type", "error");
            ret.put("msg", "请填写正确的订单信息!");
            return ret;
        }
        if(orderService.edit(order)<=0){
            ret.put("type", "error");
            ret.put("msg", "订单编辑失败,请联系管理员!");
            return ret;
        }
        ret.put("type", "success");
        ret.put("msg", "订单编辑成功!");
        return ret;
    }
    /**
     *添加订单状态信息
     * @param orderStatus
     * @return
     */
    @RequestMapping(value="/add_status",method=RequestMethod.POST)
    @ResponseBody
    public Map<String, String> addStatus(OrderStatus orderStatus){
        Map<String, String> ret=new HashMap<String, String>();
        if(orderStatus==null){
            ret.put("type", "error");
            ret.put("msg", "请填写正确的订单状态信息!");
            return ret;
        }
        orderStatus.setCreateTime(new Date());
        if(orderStatusService.add(orderStatus)<=0){
            ret.put("type", "error");
```

```
            ret.put("msg", "订单状态添加失败,请联系管理员!");
            return ret;
        }
        ret.put("type", "success");
        ret.put("msg", "订单状态添加成功!");
        return ret;
    }
}
```

5. View 层

前台页面 list.jsp 文件的源代码省略。

5.4.3 在线下单

1. 持久化层

Order.java 和 OrderStatus.java 文件的源代码在订单管理模块中已经列出。

2. DAO 层

OrderDao.java、OrderStatusDao.java、OrderMapper.xml 和 OrderStatusMapper.xml 文件的源代码在订单管理模块中已经列出。

3. Service 层

OrderService.java、OrderServiceImpl.java、OrderStatusService.java 和 OrderStatusServiceImpl.java 文件的源代码在订单管理模块中已经列出。

4. Controller 层

HomeOrderController.java 文件的源代码如下:

```java
package com.ischoolbar.programmer.controller.home;
import java.util.Date;
import java.util.HashMap;
import java.util.List;
import java.util.Map;
import org.apache.commons.lang.StringUtils;
import org.springframework.beans.factory.annotation.Autowired;
import org.springframework.stereotype.Controller;
import org.springframework.web.bind.annotation.RequestMapping;
import org.springframework.web.bind.annotation.RequestMethod;
import org.springframework.web.bind.annotation.ResponseBody;
import org.springframework.web.servlet.ModelAndView;
import com.ischoolbar.programmer.entity.admin.Order;
import com.ischoolbar.programmer.entity.admin.OrderStatus;
import com.ischoolbar.programmer.service.admin.OrderService;
import com.ischoolbar.programmer.service.admin.OrderStatusService;
/**
 * 前台控制器
```

```java
 */
@RequestMapping("/order")
@Controller
public class HomeOrderController {
    @Autowired
    private OrderService orderService;
    @Autowired
    private OrderStatusService orderStatusService;
    /**
     * 在线下单
     * @param model
     * @return
     */
    @RequestMapping(value="/online_order")
    public ModelAndView onlineOrder(ModelAndView model){
        model.addObject("orderActive", "active");
        model.setViewName("home/order/online_order");
        return model;
    }
    /**
     * 查询订单
     * @param model
     * @return
     */
    @RequestMapping(value="/query_order")
    public ModelAndView queryOrder(ModelAndView model,String sn){
        if(!StringUtils.isEmpty(sn)){
            Map<String, Object> queryMap=new HashMap<String, Object>();
            queryMap.put("orderSn", sn);
            List<OrderStatus> findList=orderStatusService.findList(queryMap);
            model.addObject("orderStatusList", findList);
            Order findBySn=orderService.findBySn(sn);
            String msg="您的订单已提交成功,等待系统分配";
            if(findBySn==null){
                msg="该单号不存在!";
            }
            model.addObject("msg", msg);
        }
        model.addObject("orderActive", "active");
        model.setViewName("home/order/query_order");
        model.addObject("sn",sn);
        return model;
    }
    /**
```

```java
 * 提交订单信息
 * @return
 */
@RequestMapping(value="/add_order",method=RequestMethod.POST)
@ResponseBody
public Map<String, Object> addOrder(Order order){
    Map<String, Object> ret=new HashMap<String, Object>();
    if(order==null){
        ret.put("type", "error");
        ret.put("msg","请填写正确的订单信息");
        return ret;
    }
    String sn=System.currentTimeMillis()+"";
    order.setSn(sn);
    order.setCreateTime(new Date());
    int add=orderService.add(order);
    if(add<0){
        ret.put("type","error");
        ret.put("msg","订单提交失败,请联系管理员");
        return ret;
    }
    ret.put("type", "success");
    ret.put("msg", "add success");
    ret.put("sn", sn);
    return ret;
}
}
```

5. View 层

online_order.jsp 页面的源代码如下:

```jsp
<%@page language="java" contentType="text/html; charset=
    UTF-8" pageEncoding="UTF-8"%>
<%@taglib prefix="c" uri="http://java.sun.com/jsp/jstl/core"%>
<%@taglib prefix="fmt" uri="http://java.sun.com/jsp/jstl/fmt"%>
<%@include file="../common/header.jsp" %>
<style>
input,.required{
    height: 40px;
    padding: 10px;
    width: 100%;
    border: 1px solid #CCCCCC;
}
</style>
<body>
```

```jsp
<%@ include file="../common/navbar.jsp"%>
    <%@ include file="../common/slide.jsp"%>
    <!-- aboupg-->
        <div class="sec aboutpg container">
            <div class="pg-nav col-sm-3">
                <div class="tit-ol">
                    <p>在线下单</p>
                </div>
                <ul>
                    <li><a href="../order/online_order">立即下单</a></li>
                    <li><a href="../order/query_order">查询订单</a></li>
                    <li><a href="../index/problem">常见问题</a></li>
                </ul>
                <div class="tit-co">
                    <p>联系我们</p>
                </div>
                <ul>
                    <li><a href="../index/contact_us">联系我们</a></li>
                </ul>
            </div>
            <div class="col-sm-9 introduce">
                <section class="title">
                    <h1>
                        在线下单
                        <span>ORDER ONLINE</span>
                    </h1>
                </section>
                <form id="order-form">
                <div class="onlinepg con-pad">
                    <div>
                        <p style="font-size:18px;font-weight:bold;color:#00389C;">
                            发货人信息</p>
                        <ul class="row">
                            <li class="col-sm-6 col-xs-12">
                                <p><i>*</i>发货人:</p>
                                <input type="text" style="" class="required"name=
                                    "sender" msg="请填写发货人姓名"/>
                            </li>
                            <li class="col-sm-6 col-xs-12">
                                <p><i>*</i>手机</p>
                                <input type="text" class="required"name=
                                    "senderMobile" msg="请填写发货人手机"
                                    maxlength="11"/>
                            </li>
```

```html
            <li class="col-sm-6 col-xs-12">
                <p>固定电话</p>
                <input type="text" name="senderTel"/>
            </li>
            <li class="col-sm-6 col-xs-12">
                <p><i>*</i>发货地址</p>
                <input type="text" class="required" name=
                    "senderAddress" msg="请填写发货人地址"/>
            </li>
        </ul>
    </div>
    <div>
        <p>收货人信息</p>
        <ul class="row">
            <li class="col-sm-6 col-xs-12">
                <p><i>*</i>收货人:</p>
                <input type="text" class="required" name="reciever"
                    msg="请填写收货人姓名"/>
            </li>
            <li class="col-sm-6 col-xs-12">
                <p><i>*</i>手机</p>
                <input type="text" class="required" name=
                    "recieverMobile" msg="请填写收货人手机"
                    maxlength="11"/>
            </li>
            <li class="col-sm-6 col-xs-12">
                <p>固定电话</p>
                <input type="text" name="recieverTel"/>
            </li>
            <li class="col-sm-6 col-xs-12">
                <p><i>* </i>收货地址</p>
                <input type="text" class="required" name=
                    "recieverAddress" msg="请填写收货人地址"/>
            </li>
        </ul>
    </div>
    <div>
        <p>货物信息</p>
        <ul class="row">
            <li class="col-sm-6 col-xs-12">
                <p><i>*</i>货物名称:</p>
                <input type="text" class="required" name=
                    "goodsName" msg="请填写货物名称"/>
            </li>
```

```html
            <li class="col-sm-6 col-xs-12">
                <p>申报价格(>100元):</p>
                <input type="text" class="required" name=
                    "goodsPrice" msg="请填写申报价格"/>
            </li>
            <li class="col-sm-6 col-xs-12">
                <p>货物重量:</p>
                <input type="text" class="required" name=
                    "goodsWeight" msg="请填写货物重量"/>
            </li>
            <li class="col-sm-6 col-xs-12">
                <p>货物体积:</p>
                <input type="text" class="required" name=
                    "goodsVolum" msg="请填写货物体积"/>
            </li>
            <li class="sm">
                <p>注:我们的工作人员会在接货时重新称重,
                    此估算仅供参考。</p>
            </li>
        </ul>
    </div>
    <div>
        <p>备注信息</p>
        <textarea name="remark"></textarea>
    </div>
    <p class="mes">注:我们的工作人员在收到发货请求时会主动联系,
        请注意接听电话。</p>
    <input type="button" id="submit-btn" value="立即下单"/>
</div>
</form>
        </div>
    </div>
<%@include file="../common/footer.jsp"%>
<script>
$(document).ready(function(){
    $("input").css({height:'40px'});
    $("#submit-btn").click(function(){
        var flag=true;
        $(".required").each(function(i,e){
            if($(e).val()==''){
                alert($(e).attr('msg'));
                flag=false;
                return false;
            }
```

```
            });
            if(flag){
                $.ajax({
                    url:'add_order',
                    type:'post',
                    data:$("#order-form").serialize(),
                    dataType:'json',
                    success:function(rst){
                        if(rst.type=='success'){
                            alert('下单成功,快递单号:'+rst.sn+'请牢记单号,以便查询
                                物流信息!');
                            window.location.href='query_order?sn='+rst.sn;
                        }else{
                            alert(rst.msg);
                        }
                    }
                });
            }
        });
    });
</script>
</body>
```

5.4.4 查询订单

1. 持久化层

Order.java 和 OrderStatus.java 文件的源代码在订单管理模块中已经列出。

2. DAO 层

OrderDao.java、OrderStatusDao.java、OrderMapper.xml 和 OrderStatusMapper.xml 文件的源代码在订单管理模块中已经列出。

3. Service 层

OrderService.java、OrderServiceImpl.java、OrderStatusService.java 和 OrderStatus-ServiceImpl.java 文件的源代码在订单管理模块中已经列出。

4. Controller 层

HomeOrderController.java 文件的源代码在在线下单模块中已经列出。

5. View 层

online_query.jsp 页面的源代码如下:

```
<%@page language="java" contentType="text/html;
charset=UTF-8" pageEncoding="UTF-8"%>
<%@taglib prefix="c" uri="http://java.sun.com/jsp/jstl/core"%>
<%@taglib prefix="fmt" uri="http://java.sun.com/jsp/jstl/fmt"%>
```

```jsp
<%@ include file="../common/header.jsp"%>
<link rel="stylesheet" type="text/css" href="../resources/home/css/order-style.css"/>
<body>
<%@ include file="../common/navbar.jsp"%>
    <%@ include file="../common/slide.jsp"%>
    <!-- aboupg-->
        <div class="sec aboutpg container">
            <div class="pg-nav col-sm-3">
                <div class="tit-ol">
                    <p>在线下单</p>
                </div>
                <ul>
                    <li><a href="../order/online_order">立即下单</a></li>
                    <li><a href="../order/query_order">查询订单</a></li>
                    <li><a href="../index/problem">常见问题</a></li>
                </ul>
                <div class="tit-co">
                    <p>联系我们</p>
                </div>
                <ul>
                    <li><a href="../index/contact_us">联系我们</a></li>
                </ul>
            </div>
            <div class="col-sm-9 introduce">
                <section class="title">
                    <h1>
                        查询物流订单
                        <span>QUERY ORDER ONLINE</span>
                    </h1>
                </section>
                <div class="onlinepg con-pad">
                    <div>
                        <form action="query_order" id="query_order_form">
                        <ul class="row">
                            <li class="col-sm-9 col-xs-12">
                                <input type="text" placeholder=
                                "请输入快递单号进行查询"
                                style="width:50%;"
                                name="sn" value="${sn }"/>
                                <input type="button" value="立即查询"
                                onclick="javascript:document.getElementById(
                                'query_order_form').submit();"
                                style="height:30px;"/>
```

```
                </li>
            </ul>
        </form>
    </div>
    <div class="aui-timeLine b-line">
        <ul class="aui-timeLine-content">
            <c:choose>
            <c:when test="${empty orderStatusList}">
            <c:if test="${not empty sn }">
            <li class="aui-timeLine-content-item">
                <em class="aui-timeLine-content-icon"></em>
                <p>${msg}</p>
            </li>
            </c:if>
            </c:when>
            <c:otherwise>
            <c:forEach items="${orderStatusList}" var="orderStatus">
            <li class="aui-timeLine-content-item">
                <em class="aui-timeLine-content-icon"></em>
                <p>${orderStatus.content }</p>
                <p style="margin-top: 10px;"><fmt:formatDate value=
                    "${orderStatus.createTime}" pattern=
                    "yyyy-MM-dd hh:mm:ss"/></p>
            </li>
            </c:forEach>
            </c:otherwise>
            </c:choose>

        </ul>
    </div>
    </div>
  </div>
 </div>
<%@ include file="../common/footer.jsp"%>
</body>
```

5.5 项目搭建

物流快递管理系统使用 eclipse 工具和 MySQL 数据库开发,系统项目结构如图 5-13 至图 5-16 所示。

```
wuliu
├── 部署描述符:
├── Java Resources
│   └── src
│       ├── com.ischoolbar.programmer.controller.admin
│       │   ├── AdminOrderController.java
│       │   ├── LeaveMsgController.java
│       │   ├── LogController.java
│       │   ├── MenuController.java
│       │   ├── NewsCategoryController.java
│       │   ├── NewsController.java
│       │   ├── RoleController.java
│       │   ├── SystemController.java
│       │   └── UserController.java
│       ├── com.ischoolbar.programmer.controller.home
│       │   ├── HomeOrderController.java
│       │   └── IndexController.java
│       └── com.ischoolbar.programmer.dao.admin
│           ├── AuthorityDao.java
│           ├── LeaveMsgDao.java
│           ├── LogDao.java
│           ├── MenuDao.java
│           ├── NewsCategoryDao.java
│           ├── NewsDao.java
│           ├── OrderDao.java
│           ├── OrderStatusDao.java
│           ├── RoleDao.java
│           └── UserDao.java
```

图 5-13 系统项目结构图 1

```
com.ischoolbar.programmer.entity.admin
├── Authority.java
├── LeaveMsg.java
├── Log.java
├── Menu.java
├── News.java
├── NewsCategory.java
├── Order.java
├── OrderStatus.java
├── Role.java
└── User.java
com.ischoolbar.programmer.interceptor.admin
└── LoginInterceptor.java
com.ischoolbar.programmer.page.admin
└── Page.java
com.ischoolbar.programmer.service.admin
├── AuthorityService.java
├── LeaveMsgService.java
├── LogService.java
├── MenuService.java
├── NewsCategoryService.java
├── NewsService.java
├── OrderService.java
├── OrderStatusService.java
├── RoleService.java
└── UserService.java
com.ischoolbar.programmer.service.admin.impl
com.ischoolbar.programmer.util
config
config.mybatis
config.mybatis.mapper.admin
config.spring
config.springmvc
```

图 5-14 系统项目结构图 2

```
├── build
└── WebContent
    ├── META-INF
    ├── resources
    └── WEB-INF
        ├── errors
        │   ├── 404.jsp
        │   └── 500.jsp
        ├── lib
        └── views
            ├── common
            │   ├── footer.jsp
            │   ├── header.jsp
            │   └── menus.jsp
            ├── home
            │   ├── common
            │   ├── index
            │   └── order
            ├── leave_msg
            │   └── list.jsp
            └── log
                └── list.jsp
```

图 5-15 系统项目结构图 3

```
├── menu
│   └── list.jsp
├── news
│   ├── add.jsp
│   ├── edit.jsp
│   └── list.jsp
├── news_category
│   └── list.jsp
├── order
│   └── list.jsp
├── role
│   └── list.jsp
├── system
│   ├── edit_password.jsp
│   ├── index.jsp
│   ├── login.jsp
│   └── welcome.jsp
├── user
│   └── list.jsp
├── web.xml
└── index.jsp
```

图 5-16 系统项目结构图 4

第6章 旅馆住宿管理系统

6.1 需求分析

6.1.1 系统概述

传统的手工客房信息管理过程复杂、烦琐,执行效率低,并且容易出错。为了提高工作效率,减少工作中的错误,我们开发了旅馆住宿管理系统。该系统可以让前台客服人员通过计算机操作进行酒店客房管理,为用户节省时间和人力,能更全面、有效地掌握酒店的基本情况,及时获取最新信息。

旅馆住宿管理系统主要以提高酒店客房服务的速度、精度,改善客户服务的亲和度,以减少工作差错为目标,减少各项资金支出,提高管理质量,从而为旅馆经营上档次创造条件。旅馆住宿流程如图6-1所示。

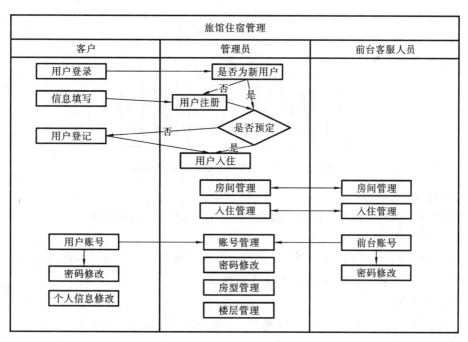

图6-1 旅馆住宿流程

6.1.2 功能需求描述

下面对旅馆住宿管理系统的功能进行分析,该系统用例图如图6-2所示。

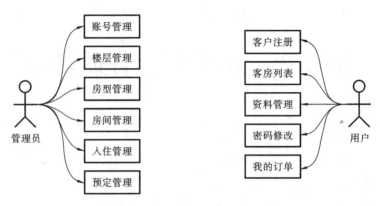

图 6-2 旅馆住宿管理系统用例图

6.2 总体设计

6.2.1 系统总体功能结构

旅馆住宿管理系统的功能模块图如图 6-3 所示。

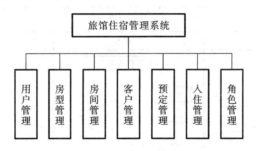

图 6-3 旅馆住宿管理系统的功能模块图

（1）用户管理。

用户管理主要对使用系统的员工进行信息修改、入职和辞退等操作。

（2）房型管理。

房型管理可以对房间类型、名称、状态、价格、床位数量等信息进行条件搜索，并且可以对相应的房型信息进行增加、修改或删除等操作。

（3）房间管理。

房间管理可以实现房间类型、房间编号、所属楼层、状态等信息的条件搜索，并且可以对相应的房间信息进行编辑和修改。

（4）客户管理。

客户管理可以实现客户登录、注册和个人信息管理功能。客户需先注册才能登录，客户需填写用户姓名、密码、确认密码和手机号。登录时需输入其用户名和密码，登录后可进行房间预定。

(5) 预定管理。

预定管理可以实现新增预定订单和编辑预定订单功能,可以根据预定的姓名、身份证号、手机号等信息进行订单查询。

(6) 入住管理。

入住管理可对客户进行登记入住、编辑入住和登记退房等操作。

(7) 角色管理。

角色管理也可称为权限管理,主要是管理员授予前台客服人员权限后可以访问后台的部分功能,以接待客户。

6.2.2 总体架构

旅馆住宿管理系统采用 SSM 框架设计,系统结构请参考第 5 章的内容。

6.2.3 数据库设计

根据 E-R 模型和数据库需求分析,旅馆住宿管理系统共涉及 7 个数据表。其中包括用户表、房间表、客户表、预订单表 4 个基本表,同时不同的登录用户要实现不同的功能,因此还需要角色权限表、入住单表,还引入了房间类型表,用于执行简单的操作。

旅馆住宿管理系统的数据表如表 6-1 至表 6-7 所示。

(1) 用户表(user):用于存储管理员信息(见表 6-1)。

表 6-1　用户表

字段名称	数据类型	主键	是否空	说明
id	int(11)	Y	N	用户 ID
username	varchar(32)	N	N	用户名
password	varchar(32)	N	N	密码
roleId	int(11)	N	N	角色 ID
sex	int(1)	N	N	性别
age	int(3)	N	N	年龄
photo	varchar(128)	N	Y	用户照片
address	varchar(128)	N	Y	地址

(2) 客户表(account):用于存储用户信息(见表 6-2)。

表 6-2　客户表

字段名称	数据类型	主键	是否空	说明
id	int(11)	Y	N	客户 ID
name	varchar(32)	N	N	客户名
password	varchar(32)	N	N	密码
realName	varchar(32)	N	Y	真实姓名

续表

字段名称	数据类型	主键	是否空	说明
idCard	varchar(32)	N	Y	身份证
mobile	varchar(16)	N	Y	电话号码
address	varchar(128)	N	Y	地址
status	int(1)	N	N	状态

（3）房间类型表（root_type）：用于存储房间类型信息（见表6-3）。

表6-3 房间类型表

字段名称	数据类型	主键	是否空	说明
id	int(11)	Y	N	主键，房间类型ID
name	varchar(64)	N	N	类型名称
photo	varchar(128)	N	Y	房间类型图片
price	float(8,2)	N	N	价格
liveNum	int(2)	N	N	可入住数
bedNum	int(5)	N	N	床位数
roomNum	int(2)	N	N	房间数
availableNum	int(5)	N	N	可用房间数
bookNum	int(5)	N	N	预定数
livedNum	int(5)	N	N	已经入住数
status	int(1)	N	N	状态
remark	varchar(256)	N	Y	备注信息

（4）房间表（room）：用于存储房间信息（见表6-4）。

表6-4 房间表

字段名称	数据类型	主键	是否空	说明
id	int(11)	Y	N	主键，房间ID
photo	varchar(128)	N	Y	房间图片
sn	varchar(32)	N	N	房间序列号
roomTypeId	int(11)	N	N	房间类型ID
floorId	int(11)	N	N	房间楼层ID
status	int(11)	N	N	状态
remark	varchar(128)	N	Y	备注信息

（5）预订单表（book_order）：用于存储预订单信息（见表6-5）。

表 6-5 预订单表

字段名称	数据类型	主键	是否空	说明
id	int(11)	Y	N	主键,预订单 ID
accountId	int(11)	N	N	管理员 ID
roomTypeId	int(11)	N	N	房间类型 ID
name	varchar(32)	N	N	预定单的名称
idCard	varchar(32)	N	Y	身份证
mobile	varchar(16)	N	Y	电话号码
status	int(1)	N	N	状态
arriveDate	varchar(32)	N	Y	到达时间
leaveDate	varchar(32)	N	N	离开时间
remark	varchar(128)	N	Y	备注信息
createTime	datetime	N	N	预定单创建时间

（6）入住单表（checkin）：用于存储入住单信息（见表 6-6）。

表 6-6 入住单表

字段名称	数据类型	主键	是否空	说明
id	int(11)	Y	N	主键,入住单 ID
roomId	int(11)	N	N	房间 ID
roomTypeId	int(11)	N	N	房间类型 ID
checkinPrice	float(8,2)	N	N	入住价格
name	varchar(32)	N	N	预定单的名称
idCard	varchar(32)	N	Y	身份证
mobile	varchar(16)	N	Y	电话号码
status	int(1)	N	N	状态
arriveDate	varchar(32)	N	Y	到达时间
leaveDate	varchar(32)	N	N	离开时间
bookOrderId	int(11)	N	Y	预定的 ID
remark	varchar(128)	N	Y	备注信息
createTime	datetime	N	N	预定单创建时间

（7）角色权限表（authority）：用于存储角色权限信息（见表 6-7）。

表 6-7 角色权限表

字 段 名	数据类型	主　键	是　否　空	说　明
id	int(11)	Y	N	主键,权限 ID
roleId	int(11)	N	N	角色 ID
menuId	int(11)	N	N	菜单 ID

6.3 详细设计

6.3.1 房间管理

房间管理模块可对房型执行增加、编辑和删除操作,还可以进行房间类型、房间编号、所属楼层和状态的条件搜索,并且可以对相应的客户信息进行编辑和修改。房间管理模块的房间列表页面如图 6-4 所示,房间列表顺序图如图 6-5 所示。

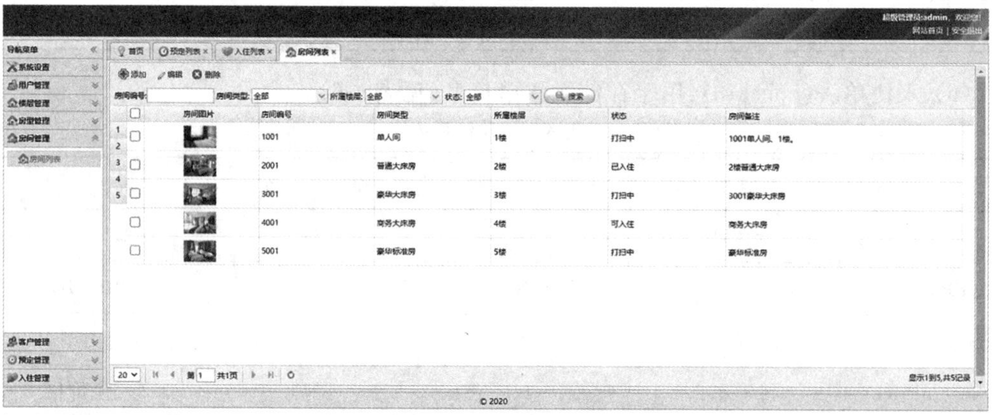

图 6-4 房间管理模块的房间列表页面

进入后台房间管理页面,通过用户键入的操作,room/list.jsp 页面将请求发送至 RoomController,RoomController 接收参数,然后调用 RoomService 的接口 RoomServiceImpl 实现功能,即调用持久层 RoomDao 接口对数据库进行操作,再将结果逐层返回至 room/list.jsp 页面并显示出来。房间管理主要是对房间进行增、删、改、查操作,且房间包含"可入住"、"已入住"、"打扫中"三个状态,由前台服务员根据客户的状态进行调整。主要由 RoomController 类实现房间管理功能。add()方法用于添加房间信息;delete()方法用于删除房间信息;edit()方法用于编辑房间信息;list()方法用于分页查询房间信息。

6.3.2 房型管理

房型管理主要是对房型进行增加、编辑和删除等操作。可以对房间类型的名称和状态进行条件搜索,可以对相应的房型信息进行修改。房型管理模块的房型列表页面如图 6-6 所示,房型列表顺序图如图 6-7 所示。

第 6 章 旅馆住宿管理系统

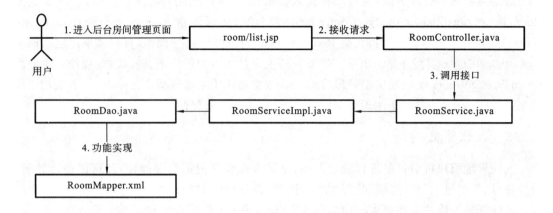

图 6-5 房间列表顺序图

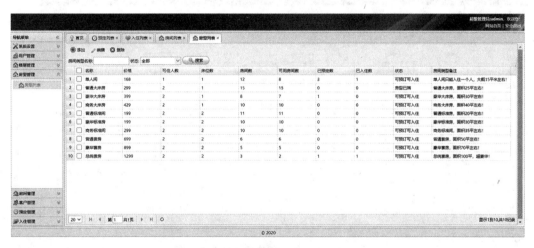

图 6-6 房型管理模块的房型列表页面

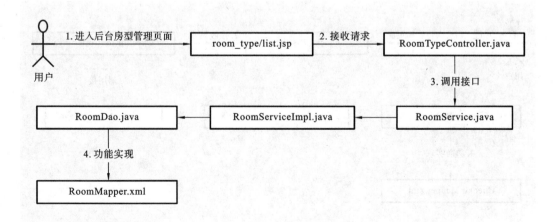

图 6-7 房型列表顺序图

进入后台房型管理页面,通过用户键入的操作,room_type/list.jsp 页面将请求发送至 RoomTypeController,RoomTypeController 接收参数,然后调用 RoomService 的接口 RoomServiceImpl 实现功能,即调用持久层 RoomDao 接口对数据库进行操作,再将结果逐层返回至 room_type/list.jsp 页面并显示出来。房型管理主要是对房间进行增、删、改、查操作。主要由 RoomTypeController 类实现房型管理功能。add()方法用于添加房型信息;delete()方法用于删除房型信息;edit()方法用于编辑房型信息;list()方法用于分页查询房型信息。

6.3.3 入住管理

入住管理模块可对订单进行登记入住、编辑入住和登记退房等操作,还可以进行姓名、身份证号、手机号、客户、房型、状态的条件搜索,并且可以对相应的客户信息进行编辑和修改。入住管理模块的入住列表页面如图 6-8 所示,用户入住顺序图如图 6-9 所示。

图 6-8 入住管理模块的入住列表页面

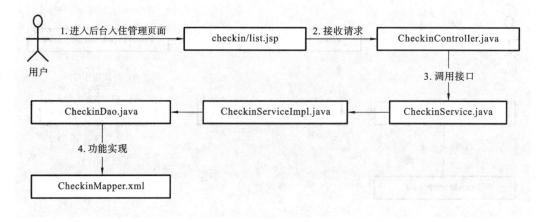

图 6-9 用户入住顺序图

进入后台入住管理页面，通过用户键入的操作，checkin/list.jsp 页面将请求发送至 CheckinController，CheckinController 接收参数，然后调用 CheckOrderService 的接口 CheckinServiceImpl 实现功能，即调用持久层 CheckinDao 接口对数据库进行操作，再将结果逐层返回至 checkin/list.jsp 页面并显示出来。入住管理系统分三个功能登记入住：登记客户预定的房间；编辑入住（根据客户留言来更改入住信息）；登记退房（客户到前台退房后，前台客服人员操作退房）。主要由 CheckinController 类来实现入住管理功能。add()方法用于添加入住信息；checkout()方法用于退房操作；edit()方法用于编辑入住信息；list()方法用于分页查询入住信息。

6.4 编码实现

本节重点介绍房型管理、房间管理和入住管理模块，公共模块和配置文件请参考第 5 章的内容。

6.4.1 房型管理

1. 持久化层

（1）RoomType.java 文件的源代码如下：

```java
package com.ischoolbar.programmer.entity;
import org.springframework.stereotype.Component;
/**
 *房间类型实体类
 */
@Component
public class RoomType {
    private Long id;                    //房间类型 id
    private String name;                //房间名称
    private String photo;               //房间类型图片
    private Float price;                //房型价格
    private Integer liveNum;            //可住人数
    private Integer bedNum;             //床位数
    private Integer roomNum;            //房间数
    private Integer avilableNum;        //可住或可预定房间数
    private Integer bookNum;            //预定数
    private Integer livedNum;           //已经入住数
    private int status;                 //房型状态,0 表示房型已满,1 表示可预定、可入住
    private String remark;              //房型备注
    //省略 getter、setter 方法
}
```

（2）Floor.java 文件的源代码如下：

```java
package com.ischoolbar.programmer.entity.admin;
```

```java
import org.springframework.stereotype.Component;
/**
 * 楼层实体类
 */
@Component
public class Floor {
    private Long id;              //楼层id
    private String name;          //楼层名称
    private String remark;        //楼层备注
    //省略getter、setter方法
}
```

2. DAO 层

(1) RoomTypeDao.java 文件的源代码如下:

```java
package com.ischoolbar.programmer.dao;
import java.util.List;
import java.util.Map;
/**
 * 房间类型 Dao
 */
import org.springframework.stereotype.Repository;
import com.ischoolbar.programmer.entity.RoomType;
@Repository
public interface RoomTypeDao {
    public int add(RoomType roomType);
    public int edit(RoomType roomType);
    public int delete(Long id);
    public List<RoomType> findList(Map<String, Object> queryMap);
    public Integer getTotal(Map<String, Object> queryMap);
    public List<RoomType> findAll();
    public RoomType find(Long id);
    public int updateNum(RoomType roomType);
}
```

(2) RoomTypeMapper.xml 文件的源代码如下:

```xml
<?xml version="1.0" encoding="UTF-8"?>
<!DOCTYPE mapper
PUBLIC "-//mybatis.org//DTD Mapper 3.0//EN"
"http://mybatis.org/dtd/mybatis-3-mapper.dtd">
<mapper namespace="com.ischoolbar.programmer.dao.admin.RoomDao">
    <!-- 执行房间插入操作 -->
    <insert id="add" parameterType="com.ischoolbar.programmer.entity.admin.Room">
        insert into room(id,photo,sn,roomTypeId,floorId,status,remark) values(null,
        #{photo},#{sn},#{roomTypeId},#{floorId},#{status},#{remark})
```

```xml
</insert>
<!-- 执行房间编辑操作 -->
<update id="edit" parameterType="com.ischoolbar.programmer.entity.admin.Room">
    update room set photo=#{photo}, sn=#{sn},roomTypeId=#{roomTypeId},floorId
        =#{floorId},status=#{status},remark=#{remark} where id=#{id}
</update>
<!-- 搜索房间信息 -->
<select id="findList" parameterType="Map" resultType=
    "com.ischoolbar.programmer.entity.admin.Room">
    select * from room where 1=1
    <if test="roomTypeId !=null">
        and roomTypeId=#{roomTypeId}
    </if>
    <if test="floorId !=null">
        and floorId=#{floorId}
    </if>
    <if test="sn !=null">
        and sn like '%${sn}%'
    </if>
    <if test="status !=null">
        and status=#{status}
    </if>
    <if test="offset !=null and pageSize !=null">
        limit #{offset},#{pageSize}
    </if>
</select>
<!-- 获取所有房间信息 -->
<select id="findAll" parameterType="Map" resultType=
    "com.ischoolbar.programmer.entity.admin.Room">
    select * from room
</select>
<!-- 获取单个房间信息 -->
<select id="find" parameterType="Long" resultType=
    "com.ischoolbar.programmer.entity.admin.Room">
    select * from room where id=#{value}
</select>
<!-- 根据房间编号获取单个房间信息 -->
<select id="findBySn" parameterType="String" resultType=
    "com.ischoolbar.programmer.entity.admin.Room">
    select * from room where sn=#{value}
</select>
<!-- 模糊搜索总条数 -->
<select id="getTotal" parameterType="Map" resultType="Integer">
    select count(*) from room where 1=1
```

```xml
            <if test="roomTypeId !=null">
                and roomTypeId=#{roomTypeId}
            </if>
            <if test="floorId !=null">
                and floorId=#{floorId}
            </if>
            <if test="sn !=null">
                and sn like '%${sn}%'
            </if>
            <if test="status !=null">
                and status=#{status}
            </if>
    </select>
    <!-- 删除房间信息 -->
    <delete id="delete" parameterType="Long">
        delete from room where id=${value}
    </delete>
</mapper>
```

(3) FloorDao.java 文件的源代码如下：

```java
package com.ischoolbar.programmer.dao.admin;
import java.util.List;
import java.util.Map;
import org.springframework.stereotype.Repository;
import com.ischoolbar.programmer.entity.admin.Floor;
/**
 * 楼层 dao
 */
@Repository
public interface FloorDao {
    public int add(Floor floor);
    public int edit(Floor floor);
    public int delete(Long id);
    public List<Floor> findList(Map<String, Object> queryMap);
    public List<Floor> findAll();
    public Integer getTotal(Map<String, Object> queryMap);
}
```

(4) FloorMapper.xml 文件的源代码如下：

```xml
<?xml version="1.0" encoding="UTF-8"?>
<!DOCTYPE mapper
PUBLIC "-//mybatis.org//DTD Mapper 3.0//EN"
"http://mybatis.org/dtd/mybatis-3-mapper.dtd">
<mapper namespace="com.ischoolbar.programmer.dao.admin.FloorDao">
```

```xml
<!-- 执行楼层插入操作 -->
<insert id="add" parameterType="com.ischoolbar.programmer.entity.admin.Floor">
    insert into floor(id,name,remark) values(null,#{name},#{remark})
</insert>
<!-- 执行楼层编辑操作 -->
<update id="edit" parameterType="com.ischoolbar.programmer.entity.admin.Floor">
    update floor set name=#{name},remark=#{remark} where id=#{id}
</update>
<!-- 搜索楼层信息 -->
<select id="findList" parameterType="Map" resultType=
    "com.ischoolbar.programmer.entity.admin.Floor">
    select * from floor
    <if test="name !=null">
        where name like '%${name}%'
    </if>
    <if test="offset !=null and pageSize !=null">
        limit #{offset},#{pageSize}
    </if>
</select>
<!-- 获取所有楼层信息 -->
<select id="findAll" parameterType="Map" resultType=
    "com.ischoolbar.programmer.entity.admin.Floor">
    select * from floor
</select>
<!-- 模糊搜索总条数 -->
<select id="getTotal" parameterType="Map" resultType="Integer">
    select count(*) from floor
    <if test="name !=null">
        where name like '%${name}%'
    </if>
</select>
<!-- 删除楼层信息 -->
<delete id="delete" parameterType="Long">
    delete from floor where id=${value}
</delete>
</mapper>
```

3. Service 层

（1）RoomTypeService.java 文件的源代码如下：

```
package com.ischoolbar.programmer.service;
/**
 *房型 service
 */
import java.util.List;
```

```java
import java.util.Map;
import org.springframework.stereotype.Service;
import com.ischoolbar.programmer.entity.RoomType;
@Service
public interface RoomTypeService {
    public int add(RoomType roomType);
    public int edit(RoomType roomType);
    public int delete(Long id);
    public List<RoomType> findList(Map<String, Object> queryMap);
    public List<RoomType> findAll();
    public Integer getTotal(Map<String, Object> queryMap);
    public RoomType find(Long id);
    public int updateNum(RoomType roomType);
}
```

(2) RoomTypeServiceImpl.java 文件的源代码如下：

```java
package com.ischoolbar.programmer.service.impl;
import java.util.List;
import java.util.Map;
import org.springframework.beans.factory.annotation.Autowired;
import org.springframework.stereotype.Service;
import com.ischoolbar.programmer.dao.RoomTypeDao;
import com.ischoolbar.programmer.entity.RoomType;
import com.ischoolbar.programmer.service.RoomTypeService;
@Service
public class RoomTypeServiceImpl implements RoomTypeService {
    @Autowired
    private RoomTypeDao roomTypeDao;
    @Override
    public int add(RoomType roomType) {
        return roomTypeDao.add(roomType);
    }
    @Override
    public int edit(RoomType roomType) {
        return roomTypeDao.edit(roomType);
    }
    @Override
    public int delete(Long id) {
        return roomTypeDao.delete(id);
    }
    @Override
    public List<RoomType> findList(Map<String, Object> queryMap) {
        return roomTypeDao.findList(queryMap);
    }
```

```
    @Override
    public List<RoomType> findAll() {
        return roomTypeDao.findAll();
    }
```

（3）FloorService.java 文件的源代码如下：

```
package com.ischoolbar.programmer.service.admin;
import java.util.List;
import java.util.Map;
import org.springframework.stereotype.Service;
import com.ischoolbar.programmer.entity.admin.Floor;
/**
 * 楼层 service
 * @author Administrator
 *
 */
@Service
public interface FloorService {
    public int add(Floor floor);
    public int edit(Floor floor);
    public int delete(Long id);
    public List<Floor> findList(Map<String, Object> queryMap);
    public List<Floor> findAll();
    public Integer getTotal(Map<String,Object> queryMap);
}
```

（4）FloorServiceImpl.java 文件的源代码如下：

```
package com.ischoolbar.programmer.service.admin.impl;
/**
 * 楼层 service 实现类
 */
import java.util.List;
import java.util.Map;
import org.springframework.beans.factory.annotation.Autowired;
import org.springframework.stereotype.Service;
import com.ischoolbar.programmer.dao.admin.FloorDao;
import com.ischoolbar.programmer.entity.admin.Floor;
import com.ischoolbar.programmer.service.admin.FloorService;
@Service
public class FloorServiceImpl implements FloorService {
    @Autowired
    private FloorDao floorDao;
    @Override
    public int add(Floor floor) {
```

```java
        return floorDao.add(floor);
    }
    @Override
    public int edit(Floor floor) {
        return floorDao.edit(floor);
    }
    @Override
    public int delete(Long id) {
        return floorDao.delete(id);
    }
    @Override
    public List<Floor> findList(Map<String, Object> queryMap) {
        return floorDao.findList(queryMap);
    }
    @Override
    public List<Floor> findAll() {
        return floorDao.findAll();
    }
    @Override
    public Integer getTotal(Map<String, Object> queryMap) {
        return floorDao.getTotal(queryMap);
    }
}
```

4. Controller 层

RoomTypeController.java 文件的源代码如下：

```java
package com.ischoolbar.programmer.controller.admin;
import java.util.HashMap;
import java.util.Map;
import org.apache.commons.lang.StringUtils;
import org.springframework.beans.factory.annotation.Autowired;
import org.springframework.stereotype.Controller;
import org.springframework.web.bind.annotation.RequestMapping;
import org.springframework.web.bind.annotation.RequestMethod;
import org.springframework.web.bind.annotation.RequestParam;
import org.springframework.web.bind.annotation.ResponseBody;
import org.springframework.web.servlet.ModelAndView;
import com.ischoolbar.programmer.entity.RoomType;
import com.ischoolbar.programmer.entity.admin.Floor;
import com.ischoolbar.programmer.page.admin.Page;
import com.ischoolbar.programmer.service.RoomTypeService;
import com.ischoolbar.programmer.service.admin.FloorService;
/**
 * 房间类型管理后台控制器
```

```java
 * @author Administrator
 *
 */
@RequestMapping("/admin/room_type")
@Controller
public class RoomTypeController {
    @Autowired
    private RoomTypeService roomTypeService;
    /**
     *房间类型管理列表页面
     * @param model
     * @return
     */
    @RequestMapping(value="/list",method=RequestMethod.GET)
    public ModelAndView list(ModelAndView model){
        model.setViewName("room_type/list");
        return model;
    }
    /**
     *房间类型信息添加操作
     * @param roomType
     * @return
     */
    @RequestMapping(value="/add",method=RequestMethod.POST)
    @ResponseBody
    public Map<String,String> add(RoomType roomType){
        Map<String,String> ret=new HashMap<String,String>();
        if(roomType==null){
            ret.put("type", "error");
            ret.put("msg", "请填写正确的房间类型信息!");
            return ret;
        }
        if(StringUtils.isEmpty(roomType.getName())){
            ret.put("type", "error");
            ret.put("msg", "房间类型名称不能为空!");
            return ret;
        }
        roomType.setAvilableNum(roomType.getRoomNum());
        //默认房间数等于可用房间数
        roomType.setBookNum(0);        //设置预定数0
        roomType.setLivedNum(0);       //设置已入住数0
        if(roomTypeService.add(roomType)<=0){
            ret.put("type", "error");
            ret.put("msg", "添加失败,请联系管理员!");
```

```java
            return ret;
        }
        ret.put("type", "success");
        ret.put("msg", "添加成功!");
        return ret;
    }
    /**
     * 房间类型信息编辑操作
     * @param roomType
     * @return
     */
    @RequestMapping(value="/edit",method=RequestMethod.POST)
    @ResponseBody
    public Map<String, String> edit(RoomType roomType){
        Map<String, String> ret=new HashMap<String, String>();
        if(roomType==null){
            ret.put("type", "error");
            ret.put("msg", "请填写正确的房间类型信息!");
            return ret;
        }
        if(StringUtils.isEmpty(roomType.getName())){
            ret.put("type", "error");
            ret.put("msg", "房间类型名称不能为空!");
            return ret;
        }
        RoomType existRoomType=roomTypeService.find(roomType.getId());
        if(existRoomType==null){
            ret.put("type", "error");
            ret.put("msg", "未找到该数据!");
            return ret;
        }
        int offset=roomType.getRoomNum()-existRoomType.getRoomNum();
        roomType.setAvilableNum(existRoomType.getAvilableNum()+offset);
        if(roomType.getAvilableNum()<=0){
            roomType.setAvilableNum(0);        //没有可用房间
            roomType.setStatus(0);             //房型已满
            if(roomType.getAvilableNum()+existRoomType.getLivedNum()+
                existRoomType.getBookNum()>roomType.getRoomNum()){
                ret.put("type", "error");
                ret.put("msg", "房间数设置不合理!");
                return ret;
            }
        }
        if(roomTypeService.edit(roomType)<=0){
```

```java
            ret.put("type", "error");
            ret.put("msg", "修改失败,请联系管理员!");
            return ret;
        }
        ret.put("type", "success");
        ret.put("msg", "修改成功!");
        return ret;
    }
    /**
     *分页查询房间类型信息
     * @param name
     * @param page
     * @return
     */
    @RequestMapping(value="/list",method=RequestMethod.POST)
    @ResponseBody
    public Map<String,Object> list(
            @RequestParam(name="name",defaultValue="") String name,
            @RequestParam(name="status",required=false) Integer status,
            Page page
            ){
        Map<String,Object> ret=new HashMap<String, Object>();
        Map<String,Object> queryMap=new HashMap<String, Object>();
        queryMap.put("name", name);
        queryMap.put("status", status);
        queryMap.put("offset", page.getOffset());
        queryMap.put("pageSize", page.getRows());
        ret.put("rows", roomTypeService.findList(queryMap));
        ret.put("total", roomTypeService.getTotal(queryMap));
        return ret;
    }
    /**
     *房间类型信息删除操作
     * @param id
     * @return
     */
    @RequestMapping(value="/delete",method=RequestMethod.POST)
    @ResponseBody
    public Map<String, String> delete(Long id){
        Map<String, String> ret=new HashMap<String, String>();
        if(id==null){
            ret.put("type", "error");
            ret.put("msg", "请选择要删除的信息!");
            return ret;
```

```java
        }
        try {
            if(roomTypeService.delete(id)<=0){
                ret.put("type", "error");
                ret.put("msg", "删除失败,请联系管理员!");
                return ret;
            }
        } catch (Exception e) {
            // TODO: handle exception
            ret.put("type", "error");
            ret.put("msg","该房间类型下存在房间信息,
                请先删除该房间类型下的所有房间信息!");
            return ret;
        }
        ret.put("type", "success");
        ret.put("msg", "删除成功!");
        return ret;
    }
}
```

5. View 层

room_type/list.jsp 文件的源代码省略。

6.4.2 房间管理

1. 持久化层

Room.java 文件的源代码如下：

```java
package com.ischoolbar.programmer.entity.admin;
import org.springframework.stereotype.Component;
/**
 *房间实体类
 */
@Component
public class Room {
    private Long id;                    //房间 id
    private String photo;               //房间图片
    private String sn;                  //房间编号
    private Long roomTypeId;            //房型 id
    private Long floorId;               //所属楼层 id
    private int status;                 //房型状态,0 表示可入住,1 表示已入住,2 表示打扫中
    private String remark;              //房型备注
    //省略 getter、setter 方法
}
```

2. DAO 层

（1）RoomDao.java 文件的源代码如下：

```java
package com.ischoolbar.programmer.dao.admin;
import java.util.List;
import java.util.Map;
/**
 *房间 dao
 */
import org.springframework.stereotype.Repository;
import com.ischoolbar.programmer.entity.admin.Room;
@Repository
public interface RoomDao {
    public int add(Room room);
    public int edit(Room room);
    public int delete(Long id);
    public List<Room> findList(Map<String, Object> queryMap);
    public List<Room> findAll();
    public Integer getTotal(Map<String, Object> queryMap);
    public Room find(Long id);
    public Room findBySn(String sn);
}
```

（2）RoomMapper.xml 文件的源代码如下：

```xml
<?xml version="1.0" encoding="UTF-8"?>
<!DOCTYPE mapper
PUBLIC "-//mybatis.org//DTD Mapper 3.0//EN"
"http://mybatis.org/dtd/mybatis-3-mapper.dtd">
<mapper namespace="com.ischoolbar.programmer.dao.admin.RoomDao">
    <!-- 房间插入操作 -->
    <insert id="add" parameterType="com.ischoolbar.programmer.entity.admin.Room">
        insert into room(id,photo,sn,roomTypeId,floorId,status,remark) values(null,
            #{photo},#{sn},#{roomTypeId},#{floorId},#{status},#{remark})
    </insert>
    <!-- 房间编辑操作 -->
    <update id="edit" parameterType="com.ischoolbar.programmer.entity.admin.Room">
        update room set photo=#{photo}, sn=#{sn},roomTypeId=#{roomTypeId},
            floorId=#{floorId},status=#{status},remark=#{remark} where id=#{id}
    </update>
    <!-- 搜索房间信息 -->
    <select id="findList" parameterType="Map" resultType=
        "com.ischoolbar.programmer.entity.admin.Room">
        select * from room where 1=1
        <if test="roomTypeId != null">
```

```xml
            and roomTypeId=#{roomTypeId}
        </if>
        <if test="floorId!=null">
            and floorId=#{floorId}
        </if>
        <if test="sn!=null">
            and sn like '%${sn}%'
        </if>
        <if test="status!=null">
            and status=#{status}
        </if>
        <if test="offset!=null and pageSize!=null">
            limit #{offset},#{pageSize}
        </if>
    </select>
    <!-- 获取所有房间信息 -->
    <select id="findAll" parameterType="Map" resultType=
        "com.ischoolbar.programmer.entity.admin.Room">
        select * from room
    </select>
    <!-- 获取单个房间信息 -->
    <select id="find" parameterType="Long" resultType=
        "com.ischoolbar.programmer.entity.admin.Room">
        select * from room where id=#{value}
    </select>
    <!-- 根据房间编号获取单个房间信息 -->
    <select id="findBySn" parameterType="String" resultType=
        "com.ischoolbar.programmer.entity.admin.Room">
        select * from room where sn=#{value}
    </select>
    <!-- 模糊搜索总条数 -->
    <select id="getTotal" parameterType="Map" resultType="Integer">
        select count(*) from room where 1=1
        <if test="roomTypeId!=null">
            and roomTypeId=#{roomTypeId}
        </if>
        <if test="floorId!=null">
            and floorId=#{floorId}
        </if>
        <if test="sn!=null">
            and sn like '%${sn}%'
        </if>
        <if test="status!=null">
            and status=#{status}
```

```
            </if>
        </select>
        <!-- 删除房间信息 -->
        <delete id="delete" parameterType="Long">
            delete from room where id=${value}
        </delete>
</mapper>
```

3. Service 层

(1) RoomService.java 文件的源代码如下:

```
package com.ischoolbar.programmer.service.admin;
/**
 * 房间 service
 */
import java.util.List;
import java.util.Map;
import org.springframework.stereotype.Service;
import com.ischoolbar.programmer.entity.admin.Room;
@Service
public interface RoomService {
    public int add(Room room);
    public int edit(Room room);
    public int delete(Long id);
    public List<Room> findList(Map<String, Object> queryMap);
    public List<Room> findAll();
    public Integer getTotal(Map<String, Object> queryMap);
    public Room find(Long id);
    public Room findBySn(String sn);
}
```

(2) RoomServiceImpl.java 文件的源代码如下:

```
package com.ischoolbar.programmer.service.admin.impl;
import java.util.List;
import java.util.Map;
import org.springframework.beans.factory.annotation.Autowired;
import org.springframework.stereotype.Service;
import com.ischoolbar.programmer.dao.admin.RoomDao;
import com.ischoolbar.programmer.entity.admin.Room;
import com.ischoolbar.programmer.service.admin.RoomService;
@Service
public class RoomServiceImpl implements RoomService {
    @Autowired
    private RoomDao roomDao;
    @Override
```

```java
    public int add(Room room) {
        return roomDao.add(room);
    }
    @Override
    public int edit(Room room) {
        return roomDao.edit(room);
    }
    @Override
    public int delete(Long id) {
            return roomDao.delete(id);
    }
    @Override
    public List<Room> findList(Map<String, Object> queryMap) {
        return roomDao.findList(queryMap);
    }
    @Override
    public List<Room> findAll() {
        return roomDao.findAll();
    }
    @Override
    public Integer getTotal(Map<String, Object> queryMap) {
        return roomDao.getTotal(queryMap);
    }
    @Override
    public Room find(Long id) {
        return roomDao.find(id);
    }
    @Override
    public Room findBySn(String sn) {
        return roomDao.findBySn(sn);
    }
}
```

4. Controller 层

RoomController.java 文件的源代码如下：

```java
package com.ischoolbar.programmer.controller.admin;
import java.util.HashMap;
import java.util.Map;
import org.apache.commons.lang.StringUtils;
import org.springframework.beans.factory.annotation.Autowired;
import org.springframework.stereotype.Controller;
import org.springframework.web.bind.annotation.RequestMapping;
import org.springframework.web.bind.annotation.RequestMethod;
import org.springframework.web.bind.annotation.RequestParam;
```

```java
import org.springframework.web.bind.annotation.ResponseBody;
import org.springframework.web.servlet.ModelAndView;
import com.ischoolbar.programmer.entity.admin.Room;
import com.ischoolbar.programmer.page.admin.Page;
import com.ischoolbar.programmer.service.RoomTypeService;
import com.ischoolbar.programmer.service.admin.FloorService;
import com.ischoolbar.programmer.service.admin.RoomService;
/**
 * 房间管理后台控制器
 * @author Administrator
 *
 */
@RequestMapping("/admin/room")
@Controller
public class RoomController {
    @Autowired
    private RoomTypeService roomTypeService;
    @Autowired
    private RoomService roomService;
    @Autowired
    private FloorService floorService;
    /**
     * 房间管理列表页面
     * @param model
     * @return
     */
    @RequestMapping(value="/list",method=RequestMethod.GET)
    public ModelAndView list(ModelAndView model){
        model.addObject("roomTypeList", roomTypeService.findAll());
        model.addObject("floorList", floorService.findAll());
        model.setViewName("room/list");
        return model;
    }
    /**
     * 房间信息添加操作
     * @param roomType
     * @return
     */
    @RequestMapping(value="/add",method=RequestMethod.POST)
    @ResponseBody
    public Map<String, String> add(Room room){
        Map<String, String> ret=new HashMap<String, String>();
        if(room==null){
            ret.put("type", "error");
```

```java
            ret.put("msg", "请填写正确的房间信息!");
            return ret;
        }
        if(StringUtils.isEmpty(room.getSn())){
            ret.put("type", "error");
            ret.put("msg", "房间编号不能为空!");
            return ret;
        }
        if(room.getRoomTypeId()==null){
            ret.put("type", "error");
            ret.put("msg", "请选择房间类型!");
            return ret;
        }
        if(room.getFloorId()==null){
            ret.put("type", "error");
            ret.put("msg", "请选择房间所属楼层!");
            return ret;
        }
        if(isExist(room.getSn(), 0l)){
            ret.put("type", "error");
            ret.put("msg", "该房间编号已经存在!");
            return ret;
        }
        if(roomService.add(room) <=0){
            ret.put("type", "error");
            ret.put("msg", "添加失败,请联系管理员!");
            return ret;
        }
        ret.put("type", "success");
        ret.put("msg", "添加成功!");
        return ret;
    }
    /**
     * 房间信息编辑操作
     * @param roomType
     * @return
     */
    @RequestMapping(value="/edit",method=RequestMethod.POST)
    @ResponseBody
    public Map<String, String> edit(Room room){
        Map<String, String> ret= new HashMap<String, String>();
        if(room==null){
            ret.put("type", "error");
            ret.put("msg", "请填写正确的房间信息!");
```

```java
            return ret;
        }
        if(StringUtils.isEmpty(room.getSn())){
            ret.put("type", "error");
            ret.put("msg", "房间编号不能为空!");
            return ret;
        }
        if(room.getRoomTypeId()==null){
            ret.put("type", "error");
            ret.put("msg", "请选择房间类型!");
            return ret;
        }
        if(room.getFloorId()==null){
            ret.put("type", "error");
            ret.put("msg", "请选择房间所属楼层!");
            return ret;
        }
        if(isExist(room.getSn(), room.getId())){
            ret.put("type", "error");
            ret.put("msg", "该房间编号已经存在!");
            return ret;
        }
        if(roomService.edit(room) <=0){
            ret.put("type", "error");
            ret.put("msg", "修改失败,请联系管理员!");
            return ret;
        }
        ret.put("type", "success");
        ret.put("msg", "修改成功!");
        return ret;
    }
    /**
     *分页查询房间信息
     * @param sn
     * @param page
     * @return
     */
    @RequestMapping(value="/list",method=RequestMethod.POST)
    @ResponseBody
    public Map<String,Object> list(
            @RequestParam(name="sn",defaultValue="") String sn,
            @RequestParam(name="status",required=false) Integer status,
            @RequestParam(name="roomTypeId",required=false) Long roomTypeId,
            @RequestParam(name="floorId",required=false) Long floorId,
```

```java
            Page page
            ){
        Map<String,Object> ret=new HashMap<String, Object>();
        Map<String,Object> queryMap=new HashMap<String, Object>();
        queryMap.put("sn", sn);
        queryMap.put("status", status);
        queryMap.put("roomTypeId", roomTypeId);
        queryMap.put("floorId", floorId);
        queryMap.put("offset", page.getOffset());
        queryMap.put("pageSize", page.getRows());
        ret.put("rows", roomService.findList(queryMap));
        ret.put("total", roomService.getTotal(queryMap));
        return ret;
    }
    /**
     * 房间信息删除操作
     * @param id
     * @return
     */
    @RequestMapping(value="/delete",method=RequestMethod.POST)
    @ResponseBody
    public Map<String, String> delete(Long id){
        Map<String, String> ret=new HashMap<String, String>();
        if(id==null){
            ret.put("type", "error");
            ret.put("msg", "请选择要删除的信息!");
            return ret;
        }
        try {
            if(roomService.delete(id) <=0){
                ret.put("type", "error");
                ret.put("msg", "删除失败,请联系管理员!");
                return ret;
            }
        } catch (Exception e) {
            ret.put("type", "error");
            ret.put("msg","该房间下存在订单信息,请先删除该房间下的所有订单信息!");
            return ret;
        }
        ret.put("type", "success");
        ret.put("msg", "删除成功!");
        return ret;
    }
```

```java
/**
 *判断房间编号是否存在
 *@param sn
 *@param id
 *@return
 */
private boolean isExist(String sn,Long id){
    Room findBySn= roomService.findBySn(sn);
    if(findBySn==null)return false;
    if(findBySn.getId().longValue()==id.longValue())return false;
    return true;
}
}
```

5．View 层

room/list.jsp 页面文件的源代码省略。

6.4.3 入住管理

1．持久化层

Checkin.java 文件的源代码如下：

```java
package com.ischoolbar.programmer.entity.admin;
import java.util.Date;
import org.springframework.stereotype.Component;
/**
 *入住实体类
 */
@Component
public class Checkin {
    private Long id;                        //入住 id
    private Long roomId;                    //房间 id
    private Long roomTypeId;                //房型 id
    private Float checkinPrice;             //入住价格
    private String name;                    //入住者姓名
    private String idCard;                  //身份证号
    private String mobile;                  //手机号
    private int status;                     //状态:0 表示入住中,1 表示已结算离店
    private String arriveDate;              //入住日期
    private String leaveDate;               //离店日期
    private Date createTime;                //创建时间
    private Long bookOrderId;               //预定订单 id,可为空
    private String remark;
    //省略 getter、setter 方法
}
```

2. DAO 层

（1）CheckinDao.java 文件的源代码如下：

```java
package com.ischoolbar.programmer.dao.admin;
import java.util.List;
import java.util.Map;
/**
 * 入住 dao
 */
import org.springframework.stereotype.Repository;
import com.ischoolbar.programmer.entity.admin.Checkin;
@Repository
public interface CheckinDao {
    public int add(Checkin checkin);
    public int edit(Checkin checkin);
    public int delete(Long id);
    public List<Checkin> findList(Map<String, Object> queryMap);
    public Integer getTotal(Map<String, Object> queryMap);
    public Checkin find(Long id);
    public List<Map> getStatsByMonth();
    public List<Map> getStatsByDay();
}
```

（2）CheckinMapper.xml 文件的源代码如下：

```xml
<?xml version="1.0" encoding="UTF-8"?>
<!DOCTYPE mapper
PUBLIC "-//mybatis.org//DTD Mapper 3.0//EN"
"http://mybatis.org/dtd/mybatis-3-mapper.dtd">
<mapper namespace="com.ischoolbar.programmer.dao.admin.CheckinDao">
    <!-- 入住插入操作 -->
    <insert id="add" parameterType="com.ischoolbar.programmer.entity.admin.Checkin">
        insert into checkin(id,roomId,roomTypeId,checkinPrice,name,idCard,mobile,
            status,arriveDate,leaveDate,createTime,bookOrderId,remark)
            values(null,#{roomId},#{roomTypeId},#{checkinPrice},#{name},
            #{idCard},#{mobile},#{status},#{arriveDate},#{leaveDate},
            #{createTime},#{bookOrderId},#{remark})
    </insert>
    <!-- 入住编辑操作 -->
    <update id="edit" parameterType="com.ischoolbar.programmer.entity.admin.Checkin">
        update checkin set roomId=#{roomId},roomTypeId=#{roomTypeId},
            checkinPrice=#{checkinPrice},name=#{name},arriveDate=#{arriveDate},
            leaveDate=#{leaveDate},idCard=#{idCard},mobile=#{mobile},status=
            #{status},remark=#{remark} where id=#{id}
    </update>
```

```xml
<!-- 搜索入住信息 -->
<select id="findList" parameterType="Map" resultType=
    "com.ischoolbar.programmer.entity.admin.Checkin">
    select * from checkin where 1=1
    <if test="name != null">
        and name like '%${name}%'
    </if>
    <if test="status != null">
        and status=#{status}
    </if>
    <if test="roomId != null">
        and roomId=#{roomId}
    </if>
    <if test="roomTypeId != null">
        and roomTypeId=#{roomTypeId}
    </if>
    <if test="idCard != null">
        and idCard like '%${idCard}%'
    </if>
    <if test="mobile != null">
        and mobile like '%${mobile}%'
    </if>
        <if test="offset != null and pageSize != null">
        limit #{offset},#{pageSize}
    </if>
</select>
<!-- 获取单个入住信息 -->
<select id="find" parameterType="Long" resultType=
    "com.ischoolbar.programmer.entity.admin.Checkin">
    select * from checkin where id=#{value}
</select>
<!-- 模糊搜索总条数 -->
<select id="getTotal" parameterType="Map" resultType="Integer">
    select count(*) from book_order where 1=1
    <if test="name != null">
        and name like '%${name}%'
    </if>
    <if test="status != null">
        and status=#{status}
    </if>
    <if test="roomId != null">
        and roomId=#{roomId}
    </if>
    <if test="roomTypeId != null">
```

```xml
            and roomTypeId=#{roomTypeId}
        </if>
        <if test="idCard !=null">
            and idCard like '%${idCard}%'
        </if>
        <if test="mobile !=null">
            and mobile like '%${mobile}%'
        </if>
    </select>
    <!-- 删除入住信息 -->
    <delete id="delete" parameterType="Long">
        delete from checkin where id=${value}
    </delete>
    <!-- 按月获取统计信息 -->
    <select id="getStatsByMonth" resultType="Map">
        select sum(checkinPrice) as money,DATE_FORMAT(createTime,'%Y-%m')
            as stats_date from checkin GROUP BY
            DATE_FORMAT(createTime,'%Y-%m')
    </select>
    <!-- 按日获取统计信息 -->
    <select id="getStatsByDay" resultType="Map">
        select sum(checkinPrice) as money,DATE_FORMAT(createTime,'%Y-%m-%d')
            as stats_date from checkin GROUP BY
            DATE_FORMAT(createTime,'%Y-%m-%d')
    </select>
</mapper>
```

3. Service 层

(1) CheckinService.java 文件的源代码如下：

```java
package com.ischoolbar.programmer.service.admin;
/**
 * 入住 service
 */
import java.util.List;
import java.util.Map;
import org.springframework.stereotype.Service;
import com.ischoolbar.programmer.entity.admin.Checkin;
@Service
public interface CheckinService {
    public int add(Checkin checkin);
    public int edit(Checkin checkin);
    public int delete(Long id);
    public List<Checkin> findList(Map<String, Object> queryMap);
    public Integer getTotal(Map<String, Object> queryMap);
```

```java
    public Checkin find(Long id);
    public List<Map> getStatsByMonth();
    public List<Map> getStatsByDay();
}
```

(2) CheckinServiceImpl.java 文件的源代码如下：

```java
package com.ischoolbar.programmer.service.admin.impl;
/**
 * 入住管理 service 实现类
 */
import java.util.List;
import java.util.Map;
import org.springframework.beans.factory.annotation.Autowired;
import org.springframework.stereotype.Service;
import com.ischoolbar.programmer.dao.admin.CheckinDao;
import com.ischoolbar.programmer.entity.admin.Checkin;
import com.ischoolbar.programmer.service.admin.CheckinService;
@Service
public class CheckinServiceImpl implements CheckinService {
    @Autowired
    private CheckinDao checkinDao;
    @Override
    public int add(Checkin checkin) {
        return checkinDao.add(checkin);
    }
    @Override
    public int edit(Checkin checkin) {
        return checkinDao.edit(checkin);
    }
    @Override
    public int delete(Long id) {
        return checkinDao.delete(id);
    }
    @Override
    public List<Checkin> findList(Map<String, Object> queryMap) {
        return checkinDao.findList(queryMap);
    }
    @Override
    public Integer getTotal(Map<String, Object> queryMap) {
        return checkinDao.getTotal(queryMap);
    }
    @Override
    public Checkin find(Long id) {
        return checkinDao.find(id);
```

```
    }
    @Override
    public List<Map> getStatsByMonth() {
        return checkinDao.getStatsByMonth();
    }
    @Override
    public List<Map> getStatsByDay() {
        return checkinDao.getStatsByDay();
    }
}
```

4. Controller 层

CheckinController.java 文件的源代码如下：

```
package com.ischoolbar.programmer.controller.admin;
import java.util.ArrayList;
import java.util.Date;
import java.util.HashMap;
import java.util.List;
import java.util.Map;
import org.apache.commons.lang.StringUtils;
import org.springframework.beans.factory.annotation.Autowired;
import org.springframework.stereotype.Controller;
import org.springframework.web.bind.annotation.RequestMapping;
import org.springframework.web.bind.annotation.RequestMethod;
import org.springframework.web.bind.annotation.RequestParam;
import org.springframework.web.bind.annotation.ResponseBody;
import org.springframework.web.servlet.ModelAndView;
import com.ischoolbar.programmer.entity.BookOrder;
import com.ischoolbar.programmer.entity.RoomType;
import com.ischoolbar.programmer.entity.admin.Checkin;
import com.ischoolbar.programmer.entity.admin.Room;
import com.ischoolbar.programmer.page.admin.Page;
import com.ischoolbar.programmer.service.BookOrderService;
import com.ischoolbar.programmer.service.RoomTypeService;
import com.ischoolbar.programmer.service.admin.CheckinService;
import com.ischoolbar.programmer.service.admin.RoomService;
/**
 * 入住管理后台控制器
 */
@RequestMapping("/admin/checkin")
@Controller
public class CheckinController {
    @Autowired
    private RoomService roomService;
```

```java
@Autowired
private RoomTypeService roomTypeService;
@Autowired
private BookOrderService bookOrderService;
@Autowired
private CheckinService checkinService;
/**
 * 入住管理列表页面
 * @param model
 * @return
 */
@RequestMapping(value="/list",method=RequestMethod.GET)
public ModelAndView list(ModelAndView model){
    model.addObject("roomTypeList", roomTypeService.findAll());
    model.addObject("roomList", roomService.findAll());
    model.setViewName("checkin/list");
    return model;
}
/**
 * 入住信息添加操作
 * @param checkin
 * @return
 */
@RequestMapping(value="/add",method=RequestMethod.POST)
@ResponseBody
public Map<String, String> add(Checkin checkin,
        @RequestParam(name="bookOrderId",required=false) Long bookOrderId
        ){
    Map<String, String> ret=new HashMap<String, String>();
    if(checkin==null){
        ret.put("type", "error");
        ret.put("msg", "请填写正确的入住信息!");
        return ret;
    }
    if(checkin.getRoomId()==null){
        ret.put("type", "error");
        ret.put("msg", "房间不能为空!");
        return ret;
    }
    if(checkin.getRoomTypeId()==null){
        ret.put("type", "error");
        ret.put("msg", "房型不能为空!");
        return ret;
    }
```

```java
if(StringUtils.isEmpty(checkin.getName())){
    ret.put("type", "error");
    ret.put("msg", "入住联系人名称不能为空!");
    return ret;
}
if(StringUtils.isEmpty(checkin.getMobile())){
    ret.put("type", "error");
    ret.put("msg", "入住联系人手机号不能为空!");
    return ret;
}
if(StringUtils.isEmpty(checkin.getIdCard())){
    ret.put("type", "error");
    ret.put("msg", "联系人身份证号不能为空!");
    return ret;
}
if(StringUtils.isEmpty(checkin.getArriveDate())){
    ret.put("type", "error");
    ret.put("msg", "到达时间不能为空!");
    return ret;
}
if(StringUtils.isEmpty(checkin.getLeaveDate())){
    ret.put("type", "error");
    ret.put("msg", "离店时间不能为空!");
    return ret;
}
checkin.setCreateTime(new Date());
if(checkinService.add(checkin) <=0){
    ret.put("type", "error");
    ret.put("msg", "添加失败,请联系管理员!");
    return ret;
}
RoomType roomType=roomTypeService.find(checkin.getRoomTypeId());
if(bookOrderId !=null){
    //从预定来的入住单(入住既可以是直接入住,也可以是已经预定的人来入住)
    BookOrder bookOrder=bookOrderService.find(bookOrderId);
    bookOrder.setStatus(1);
    bookOrderService.edit(bookOrder);
}else{
    roomType.setAvilableNum(roomType.getAvilableNum() -1);
}
//入住成功后去修改该房型的预定数
if(roomType !=null){
    roomType.setLivedNum(roomType.getLivedNum()+1);   //入住数加1
    roomTypeService.updateNum(roomType);
```

```java
            //如果可用的房间数为 0,则设置该房型状态已满
            if(roomType.getAvilableNum()==0){
                roomType.setStatus(0);
                roomTypeService.edit(roomType);
            }
        }
        Room room=roomService.find(checkin.getRoomId());
        if(room !=null){
            //要把房间状态设置为已入住
            room.setStatus(1);
            roomService.edit(room);
        }
        ret.put("type", "success");
        ret.put("msg", "添加成功!");
        return ret;
    }
    /**
     *入住信息编辑操作
     *@param account
     *@return
     */
    @RequestMapping(value="/edit",method=RequestMethod.POST)
    @ResponseBody
    public Map<String, String> edit(Checkin checkin,
            @RequestParam(name="bookOrderId",required=false) Long bookOrderId
            ){
        Map<String, String> ret=new HashMap<String, String>();
        if(checkin==null){
            ret.put("type", "error");
            ret.put("msg", "请填写正确的入住信息!");
            return ret;
        }
        if(checkin.getRoomId()==null){
            ret.put("type", "error");
            ret.put("msg", "房间不能为空!");
            return ret;
        }
        if(checkin.getRoomTypeId()==null){
            ret.put("type", "error");
            ret.put("msg", "房型不能为空!");
            return ret;
        }
        if(StringUtils.isEmpty(checkin.getName())){
            ret.put("type", "error");
```

```java
            ret.put("msg", "入住联系人名称不能为空!");
            return ret;
        }
        if(StringUtils.isEmpty(checkin.getMobile())){
            ret.put("type", "error");
            ret.put("msg", "入住联系人手机号不能为空!");
            return ret;
        }
        if(StringUtils.isEmpty(checkin.getIdCard())){
            ret.put("type", "error");
            ret.put("msg", "联系人身份证号不能为空!");
            return ret;
        }
        if(StringUtils.isEmpty(checkin.getArriveDate())){
            ret.put("type", "error");
            ret.put("msg", "到达时间不能为空!");
            return ret;
        }
        if(StringUtils.isEmpty(checkin.getLeaveDate())){
            ret.put("type", "error");
            ret.put("msg", "离店时间不能为空!");
            return ret;
        }
        Checkin existCheckin=checkinService.find(checkin.getId());
        if(existCheckin==null){
            ret.put("type", "error");
            ret.put("msg", "请选择正确的入住信息进行编辑!");
            return ret;
        }
        if(checkinService.edit(checkin) <=0){
            ret.put("type", "error");
            ret.put("msg", "编辑失败,请联系管理员!");
            return ret;
        }
        //编辑成功之后:1表示判断房型是否发生变化,2表示判断房间是否发生变化,3表示判断
          是否是从预定订单来的信息
        //判断是否是从预定来的入住信息
        RoomType oldRoomType=roomTypeService.find(
            existCheckin.getRoomTypeId());
        RoomType newRoomType=roomTypeService.find(checkin.getRoomTypeId());
        //房型入住数不受预定订单的影响
        if(oldRoomType.getId().longValue() !=newRoomType.getId().longValue()){
            //说明房型发生了变化,原来的房型入住数恢复,新的房型入住数增加
            oldRoomType.setLivedNum(oldRoomType.getLivedNum()-1);
```

```java
            newRoomType.setLivedNum(newRoomType.getLivedNum()+1);
            if(bookOrderId==null){
                oldRoomType.setAvilableNum(oldRoomType.getAvilableNum()+1);
                newRoomType.setAvilableNum(newRoomType.getAvilableNum()-1);
            }
        }
/**
        if(bookOrderId==null){
            //表示不是从预定订单来的,此时需判断原来的入住信息是否来源于预定
            if(existCheckin.getBookOrderId()==null){
                oldRoomType.setAvilableNum(oldRoomType.getAvilableNum()+1);
                newRoomType.setAvilableNum(newRoomType.getAvilableNum()-1);
            }
            if(existCheckin.getBookOrderId() !=null){
                //表示原来的入住信息来源于预定,但是新的入住信息不是来源于预定,需恢复
                  原来的预定状态
                BookOrder oldBookOrder=bookOrderService.find(
                    existCheckin.getBookOrderId());
                oldBookOrder.setStatus(0);
                bookOrderService.edit(oldBookOrder);
                oldRoomType.setBookNum(oldRoomType.getBookNum()+1);
            }
        }
        //表示此时的订单是来源于预定
        if(bookOrderId !=null){
            //表示是从预定订单来的,此时需判断原来的入住信息是否来源于预定
            if(existCheckin.getBookOrderId() !=null){
            //表示原来的入住信息来源于预定,但是新的入住信息不是来源于预定,需恢复原来
              的预定状态
                BookOrder oldBookOrder=bookOrderService.find(
                    existCheckin.getBookOrderId());
                if(bookOrderId.longValue() !=oldBookOrder.getId().longValue()){
                    oldBookOrder.setStatus(0);
                    bookOrderService.edit(oldBookOrder);
                }
            }
            if(oldRoomType.getId().longValue() !=newRoomType.getId().longValue()){
                newRoomType.setBookNum(newRoomType.getBookNum()-1);
                
                if(existCheckin.getBookOrderId()==null){
                    oldRoomType.setAvilableNum(oldRoomType.getAvilableNum()+1);
                }else{
                    oldRoomType.setBookNum(oldRoomType.getBookNum()+1);
                }
```

```java
        }
        BookOrder newBookOrder=bookOrderService.find(bookOrderId);
        newBookOrder.setStatus(1);
        bookOrderService.edit(newBookOrder);
    }**/
        roomTypeService.updateNum(newRoomType);
        roomTypeService.updateNum(oldRoomType);
        //判断房间是否发生变化
        if(checkin.getRoomId().longValue()!=existCheckin.getRoomId().longValue()){
            //表示房间发生了变化
            Room oldRoom=roomService.find(existCheckin.getRoomId());
            Room newRoom=roomService.find(checkin.getRoomId());
            oldRoom.setStatus(0);      //原来的房间可入住
            newRoom.setStatus(1);      //现在的房间已入住
            roomService.edit(newRoom);
            roomService.edit(oldRoom);
        }
        ret.put("type", "success");
        ret.put("msg", "修改成功!");
        return ret;
}
/**
 * 分页查询入住信息
 * @param name
 * @param page
 * @return
 */
@RequestMapping(value="/list",method=RequestMethod.POST)
@ResponseBody
public Map<String,Object> list(
        @RequestParam(name="name",defaultValue="") String name,
        @RequestParam(name="roomId",defaultValue="") Long roomId,
        @RequestParam(name="roomTypeId",defaultValue="") Long roomTypeId,
        @RequestParam(name="idCard",defaultValue="") String idCard,
        @RequestParam(name="mobile",defaultValue="") String mobile,
        @RequestParam(name="status",required=false) Integer status,
        Page page
        ){
    Map<String,Object> ret=new HashMap<String, Object>();
    Map<String,Object> queryMap=new HashMap<String, Object>();
    queryMap.put("name", name);
    queryMap.put("status", status);
    queryMap.put("roomId", roomId);
    queryMap.put("roomTypeId", roomTypeId);
```

```java
            queryMap.put("idCard", idCard);
            queryMap.put("mobile", mobile);
            queryMap.put("offset", page.getOffset());
            queryMap.put("pageSize", page.getRows());
            ret.put("rows", checkinService.findList(queryMap));
            ret.put("total", checkinService.getTotal(queryMap));
            return ret;
    }
    /**
     *退房操作
     *@param checkId
     *@return
     */
    @RequestMapping(value="/checkout",method=RequestMethod.POST)
    @ResponseBody
    public Map<String, String> checkout(Long checkId
            ){
        Map<String, String> ret=new HashMap<String, String>();
        if(checkId==null){
            ret.put("type", "error");
            ret.put("msg", "请选择数据!");
            return ret;
        }
        Checkin checkin=checkinService.find(checkId);
        if(checkin==null){
            ret.put("type", "error");
            ret.put("msg", "请选择正确的数据!");
            return ret;
        }
        checkin.setStatus(1);
        if(checkinService.edit(checkin)<=0){
            ret.put("type", "error");
            ret.put("msg", "退房失败,请联系管理员!");
            return ret;
        }
        //首先操作房间状态
        Room room=roomService.find(checkin.getRoomId());
        if(room !=null){
            room.setStatus(2);
            roomService.edit(room);
        }
        //其次修改房型可用数、入住数和状态
        RoomType roomType=roomTypeService.find(checkin.getRoomTypeId());
        if(roomType !=null){
```

```java
            roomType.setAvilableNum(roomType.getAvilableNum()+1);
            if(roomType.getAvilableNum()>roomType.getRoomNum()){
                roomType.setAvilableNum(roomType.getRoomNum());
            }
            roomType.setLivedNum(roomType.getLivedNum()-1);
            if(roomType.getStatus()==0){
                roomType.setStatus(1);
            }
            if(checkin.getBookOrderId()!=null){
                roomType.setBookNum(roomType.getBookNum()-1);
            }
            roomTypeService.updateNum(roomType);
            roomTypeService.edit(roomType);
        }
        //判断订单是否来自预定
        if(checkin.getBookOrderId()!=null){
            BookOrder bookOrder=bookOrderService.find(checkin.getBookOrderId());
            bookOrder.setStatus(2);
            bookOrderService.edit(bookOrder);

        }
        ret.put("type", "success");
        ret.put("msg", "退房成功!");
        return ret;
    }
    /**
     * 根据房间类型获取房间
     * @param roomTypeId
     * @return
     */
    @RequestMapping(value="/load_room_list",method=RequestMethod.POST)
    @ResponseBody
    public List<Map<String, Object>> load_room_list(Long roomTypeId){
        List<Map<String, Object>> retList=new ArrayList<Map<String,Object>>();
        Map<String, Object> queryMap=new HashMap<String, Object>();
        queryMap.put("roomTypeId", roomTypeId);
        queryMap.put("status", 0);
        queryMap.put("offset", 0);
        queryMap.put("pageSize", 999);
        List<Room> roomList=roomService.findList(queryMap);
        for(Room room:roomList){
            Map<String, Object> option=new HashMap<String, Object>();
```

```
                option.put("value", room.getId());
                option.put("text", room.getSn());
                retList.add(option);
            }
            return retList;
        }
    }
```

5. View 层

checkin/list.jsp 文件的源代码省略。

6.5 项目搭建

旅馆住宿管理系统使用 eclipse 工具和 MySQL 数据库开发，系统项目结构如图 6-10 至图 6-13 所示。

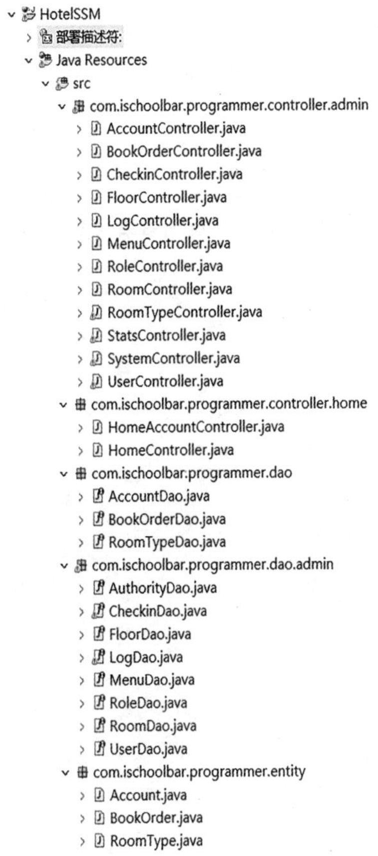

图 6-10　系统项目结构图 1

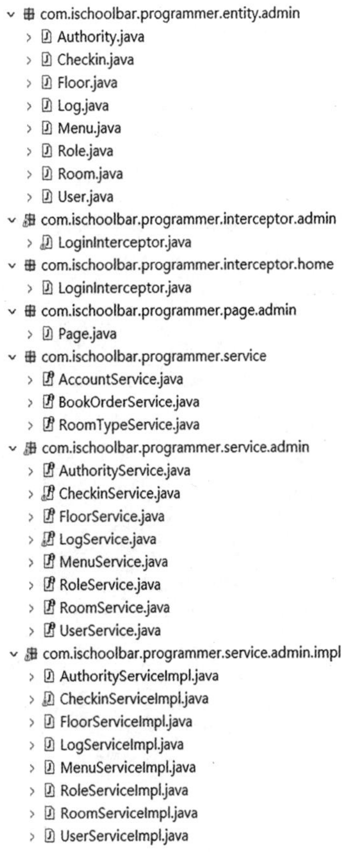

图 6-11　系统项目结构图 2

图 6-12　系统项目结构图 3　　　图 6-13　系统项目结构图 4

第 7 章 火车订票系统

7.1 需求分析

7.1.1 系统概述

随着计算机技术的发展，使用计算机提高人们的生活已经普及。铁路公司为了加强公司的信息化建设，提高公司的售票管理效率，建立了相应的火车订票系统。火车订票系统在很大程度上解决了用户订票、退票的问题，与以往只能通过线下购票相比，方便且效率更高。

火车订票系统是通过计算机实现票务信息的统一管理，以提高工作效率，让售票员售票和乘客购票更加方便。火车订票系统的目标是提供及时、准确的信息服务，加快信息检索的效率，加速票务信息实况的查询，减轻管理员制作报表和统计分析的负担。

7.1.2 功能需求描述

通过对传统火车购票流程的分析，我们设计了完整的网上火车订票系统，该系统能够轻松地让乘客查询及购买火车票。火车订票系统必须具备如下功能。

（1）用户注册、登录等。
（2）用户查票、购票、个人信息管理、退票、改签等。
（3）管理员能够对车次信息、车站信息和类型等进行管理。
（4）能够对网站日常新闻发布、动态进行维护。

7.2 总体设计

7.2.1 系统总体功能结构

分析火车订票系统中的各个模块之后，可以实现不同的账号登录功能。注册用户登录后，可以使用用户账号购票、查询火车信息等。管理员账号登录后，可以使用管理权限，管理火车订票系统的各部分信息以及用户，如管理用户信息、售票订单、火车车次等。

管理员模块如图 7-1 所示。
车次信息管理：对火车车次信息进行添加、修改、删除等操作。
火车途径信息管理：对车次途经站的详情信息进行添加、修改、删除等操作。
订单信息管理：对订单信息进行处理、删除等操作。
用户信息管理：对火车乘客信息进行添加、修改、删除等操作。
注册用户模块如图 7-2 所示。

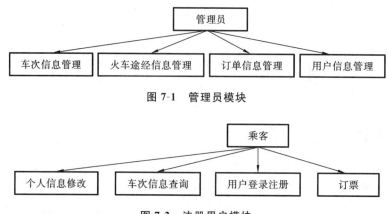

图 7-1 管理员模块

图 7-2 注册用户模块

个人信息修改：乘客可以对自己的信息进行修改，如修改自己的登录密码等。
车次信息查询：查询需要乘坐的火车车次。
用户登录注册：乘客录入信息注册账号。
订票：目前已付款的订单，可以进行退票、改签等操作。

7.2.2 总体架构

火车订票系统采用 SSM 框架设计，系统结构请参考第 5 章的内容。

7.2.3 数据库设计

（1）管理员（admin）信息：管理员 id、管理员账户（username）和密码（password）。

（2）订单（order）信息：订单 id、车辆信息 id（car_info_id）、客户 id（person_id）、改签次数（change_times）和状态（status）。

（3）车次（trips）信息：车次 id、出发地（orgin_location）、目的地（destination_location）、开始时间（start_time）、到达时间（reach_time）、跨越天数（span_days）、火车序列号（car_num）、票价（ticket_price）和票数（ticket_num）。

（4）火车途径（trips_via）信息：车次途径 id、途经站（station_name）、开始时间（start_time）、到达时间（reach_time）、火车序列号（car_num）和订单数（order_num）。

（5）用户（user）资料信息：用户 id、用户姓名（username）、用户真实姓名（true_name）、密码（password）、身份证号（id_card_nume）、电话号码（phone_num）、年龄（age）和性别（sex）。

7.3 详细设计

7.3.1 车次管理

车次管理模块对车次的起始站、目的站、出发时间、到达时间、剩余票数等信息进行添加、修改等操作。车次管理页面如图 7-3 所示，修改车次页面如图 7-4 所示。

第 7 章　火车订票系统

图 7-3　车次管理页面

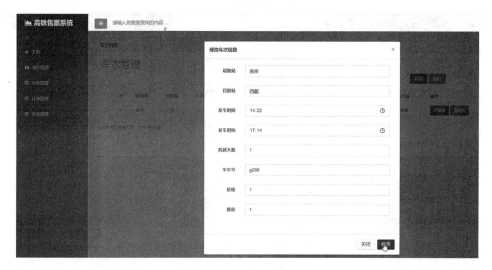

图 7-4　修改车次页面

7.3.2　订单管理

注册用户选择出发城市、到达城市及出发日期后，进入购票页面，用户订购车票后生成订单。车次选择页面如图 7-5 所示，车票查询页面如图 7-6 所示，购票确认页面如图 7-7 所

图 7-5　车次选择页面

示,下单后订单页面如图7-8所示。

图 7-6　车票查询页面

图 7-7　购票确认页面

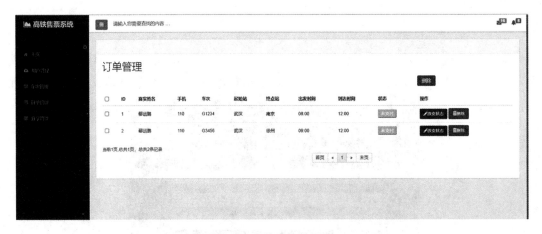

图 7-8　下单后订单页面

用户确认购票后,管理员系统管理页面会出现新订单。

7.4 编码实现

本节重点介绍车次管理和订单管理模块,公共模块和配置文件请参考第 5 章的内容。

7.4.1 车次管理

1. 持久化层

(1) Trips.java 文件的源代码如下:

```java
package com.code.rts.entity;
import org.springframework.format.annotation.DateTimeFormat;
/**
 *车次信息
 **/
public class Trips {
    private int id;
    private String orginLocation;
    private String destinationLocation;
    @DateTimeFormat(pattern="hh:mm:ss")
    private String startTime;
    @DateTimeFormat(pattern="hh:mm:ss")
    private String reachTime;
    private String spanDays;
    private String carNum;
    private Double ticketPrice;
    private int ticketNum;
    //省略 getter、setter 方法
}
```

(2) User.java 文件的源代码如下:

```java
package com.code.rts.entity;
/**
 *用户注册信息 pojo
 *用于存储用户账号信息
 */
@Getter
@Setter
public class User {
    private int id;
    private String username;
    private String password;
    private String trueName;
    private String idCardNum;
```

```java
    private String phoneNum;
    private String age;
    private String sex;
    //省略 getter、setter 方法
}
```

2. DAO 层

（1）TripsDao.java 文件的源代码如下：

```java
package com.code.rts.dao;
import com.code.rts.entity.*;
import com.github.pagehelper.Page;
import org.apache.ibatis.annotations.*;
import java.util.List;
public interface TripsDao {
    /**
     * 查询目标车票信息
     * @param trips
     * @return
     **/
    @Select("select* from trips where start_time like CONCAT('%',#{startTime},'%')"
        + " and car_num=#{carNum}")
    Trips getAimtrips(Trips trips);
    /**
     * 查询全部车票信息
     * @param trips
     * @return
     **/
    @Select("select* from trips where orgin_location like CONCAT("
        + "'%',#{orginLocation},'%') and destination_location like CONCAT("
        + "'%',#{destinationLocation},'%')")
    List<Trips> getAlltrips(Trips trips);
    /**
     * 分页查询全部车票
     * @return
     */
    Page<Trips> getAllTripsForAdmin();
    /***
     * @param id
     * @return
     */
    @Select("select * from trips where id=#{id}")
    Trips gettrips(int id);
    /**
     * 在后台添加车次
```

```java
 * @param trips
 * @return
 */
@Insert("insert into 'trips' (orgin_location, destination_location, start_time,
    reach_time, span_days, car_num, ticket_price, ticket_num) values("
    + "#{orginLocation}, #{destinationLocation}, #{startTime}, #{reachTime},"
    + "#{spanDays}, #{carNum}, #{ticketPrice}, #{ticketNum})")
int saveTrip(Trips trips);
/**
 * 订单改签后改变旧车票信息
 * @param oldId
 * @return
 **/
@Update("update trips set ticket_num=ticket_num+1 where id=#{oldId}")
int changeOldtrips(int oldId);
/**
 * 订单改签后改变新车票信息
 * @param newId
 * @return
 **/
@Update("update trips set ticket_num=ticket_num-1 where id=#{newId}")
int changeNewtrips(int newId);
/**
 * 通过id和车号查询车次信息
 * @param trips
 * @return
 */
@Select("select * from trips where id=#{id} and car_num=#{carNum}")
Trips getTripsInfoByCarInfoIdAndId(Trips trips);
/**
 * 更新trips表
 * @param trips
 * @return
 */
@Update(" <script> update trips set <if test='ticketNum !=0'> ticket_num
    =#{ticketNum}</if>"+"where id=#{id}</script> ")
int updateTrips(Trips trips);
@Update("update trips set ticket_num=ticket_num+1"+
        " where   car_num=#{carNum}" +
        "and start_time=#{startTime} and reach_time=#{reachTime}")
int refundTrips(@Param("personId") int personId, @Param("carNum") String carNum,
    @Param("startTime") String startTime, @Param("reachTime") String reachTime);
@Select("select *  from trips where car_num=#{carNum} and start_time=
    #{startTime}")
```

```java
    Trips getTripsInfoByCarNumAndStartTime(@Param("carNum") String carNum,
        @Param("startTime") String startTime);
    /**
     * 更新用户信息
     * @param trips
     * @return
     */
    int updateTripForAdmin(Trips trips);
    /**
     * 根据id删除车次
     * @param id
     * @return
     */
    Integer deleteTrip(Integer id);
}
```

（2）TripsDao.xml 文件的源代码如下：

```xml
<?xml version="1.0" encoding="UTF-8"?>
<!DOCTYPE mapper
    PUBLIC "-//mybatis.org//DTD Mapper 3.0//EN"
    "http://mybatis.org/dtd/mybatis-3-mapper.dtd">
<!-- 对应接口,写入SQL语句 -->
<mapper namespace="com.code.rts.dao.TripsDao">
    <select id="getAllTripsForAdmin" resultType="com.code.rts.entity.Trips"
        parameterType="com.github.pagehelper.Page">
        SELECT * FROM 'trips'
    </select>
    <update id="updateTripForAdmin" parameterType="com.code.rts.entity.Trips">
        UPDATE 'trips' SET
        <if test="orginLocation !=null and orginLocation !=''">
            orgin_location=#{orginLocation}
        </if>
        <if test="destinationLocation !=null and destinationLocation !=''">
            ,destination_location=#{destinationLocation}
        </if>
        <if test="startTime !=null and startTime !=''">
            ,start_time=#{startTime}
        </if>
        <if test="reachTime !=null and reachTime !=''">
            ,reach_time=#{reachTime}
        </if>
        <if test="spanDays !=null and spanDays !=''">
            ,span_days=#{spanDays}
        </if>
```

```xml
        <if test="carNum!=null and carNum!=''">
            ,car_num=#{carNum}
        </if>
        <if test="ticketPrice!=null and ticketPrice!=''">
            ,ticket_price=#{ticketPrice}
        </if>
        <if test="ticketNum!=null and ticketNum!=''">
            ,ticket_num=#{ticketNum}
        </if>
        where id=#{id}
    </update>
    <!-- 根据 id 删除用户 -->
    <delete id="deleteTrip" parameterType="Integer">
      DELETE FROM 'trips' WHERE id=#{id}
    </delete>
</mapper>
```

(3) UserDao.java 文件的源代码如下:

```java
package com.code.rts.dao;
import com.code.rts.entity.*;
import com.github.pagehelper.Page;
import org.apache.ibatis.annotations.*;
/**
 *User 表持久类
 */
public interface UserDao {
    /**
     *通过 username 查询 User 信息
     * @param username
     * @return
     */
    @Select("select * from user where username=#{username}")
    User getUserByUsername(String username);
    /**
     *注册用户信息
     * @param user
     * @return
     */
    @Insert("insert into user (username, password) values(#{username}, #{password})")
    int insertUserRegisterInfo(User user);
    /**
     *更新用户信息
     * @param user
     * @return
```

```java
     */
    int updateUser(User user);
    /**
     *更新用户账户信息
     *@param user
     *@return
     */
    int updateUserRegisterInfo(User user);
    /**
     *查询所有用户
     *@return
     */
    Page<User> getAllUsers();
    /**
     *查询所有用户的个数
     *@return
     */
    Integer getUsersCount();
    /**
     *根据 id 删除用户
     *@param id
     *@return
     */
    Integer deleteUser(Integer id);
    /**
     *在后台添加用户
     *@param user
     *@return
     */
    @Insert("insert into user (username, password, true_name, id_card_num,
        phone_num, age, sex) values(#{username}, #{password}, #{trueName},
        #{idCardNum}, #{phoneNum}, #{age}, #{sex})")
    int saveUser(User user);
}
```

（4）UserDao.xml 文件的源代码如下：

```xml
<?xml version="1.0" encoding="UTF-8"?>
<!DOCTYPE mapper
    PUBLIC "-//mybatis.org//DTD Mapper 3.0//EN"
    "http://mybatis.org/dtd/mybatis-3-mapper.dtd">
<!-- 对应接口,写入 SQL 语句 -->
<mapper namespace="com.code.rts.dao.UserDao">
    <update id="updateUserRegisterInfo" parameterType="com.code.rts.entity.User">
```

```xml
        UPDATE 'user' set
        <if test="password != null and password != ''">
            password=#{password}
        </if>
        where username=#{username}
    </update>
    <update id="updateUser" parameterType="com.code.rts.entity.User">
        UPDATE user SET
        <if test="username != null and username != ''">
            username=#{username}
        </if>
        <if test="trueName != null and trueName != ''">
            ,true_name=#{trueName}
        </if>
        <if test="idCardNum != null and idCardNum != ''">
            ,id_card_num=#{idCardNum}
        </if>
        <if test="phoneNum != null and phoneNum != ''">
            ,phone_num=#{phoneNum}
        </if>
        <if test="age != null and age != ''">
            ,age=#{age}
        </if>
        <if test="sex != null and sex != ''">
            ,sex=#{sex}
        </if>
        where id=#{id}
    </update>
    <select id="getUsersCount" resultType="Integer">
        SELECT COUNT(1)
        FROM 'user'
    </select>
    <select id="getAllUsers" resultType="com.code.rts.entity.User"
        parameterType="com.github.pagehelper.Page">
        SELECT id, username, true_name, id_card_num, phone_num, age, sex
        FROM 'user'
    </select>
    <!-- 根据id删除用户 -->
    <delete id="deleteUser" parameterType="Integer">
        DELETE FROM 'user' WHERE id=#{id}
    </delete>
</mapper>
```

3. Service 层

TripsService.java 文件的源代码如下：

```java
package com.code.rts.service;
import com.code.rts.Result.Result;
import com.code.rts.dao.OrderDao;
import com.code.rts.dao.TripsDao;
import com.code.rts.dao.UserDao;
import com.code.rts.entity.Order;
import com.code.rts.entity.Trips;
import com.code.rts.entity.User;
import com.github.pagehelper.Page;
import org.springframework.stereotype.Service;
import org.springframework.transaction.annotation.Transactional;
import javax.annotation.Resource;
import java.util.HashMap;
import java.util.List;
import java.util.Map;
@Service
public class TripsService {
    @Resource
    private OrderDao orderDao;
    @Resource
    private UserDao userDao;
    @Resource
    private TripsDao tripsDao;
    public Result getAlltrips(Trips trips) {
        Result result=new Result();
        List<Trips> tripsdata=tripsDao.getAlltrips(trips);
        if(tripsdata !=null){
            result.setMsg("Query all succeed");
            result.setData(tripsdata);
            result.setStateCode(200);
        }
        else{
            result.setMsg("Query failes,no tickets");
            result.setStateCode(404);
        }
        return result;
    }
    public Page<Trips> getAllTripsForAdmin() {
        return tripsDao.getAllTripsForAdmin();
    }
    public Result getAimtrips(Trips trips) {
        Result result=new Result();
        Trips tripsdata=tripsDao.getAimtrips(trips);
        if(tripsdata !=null){
```

```java
            result.setMsg("Query all succeed");
            result.setData(trips);
            result.setStateCode(200);
        }
        else{
            result.setMsg("Query failes,no tickets");
            result.setStateCode(404);
        }
        return result;
    }
    @Transactional
    public Result buyTicket(String username, int carInfoId, String carNum) {
        Result result=new Result();
        //获取用户个人信息的id
        User customer=userDao.getUserByUsername(username);
        if (customer.getTrueName()==null || customer.getIdCardNum()==
        null || customer.getPhoneNum()==null){
            result.setStateCode(400);
            result.setMsg("购票前请完善用户个人信息");
            result.setData(false);
            return result;
        }
        Trips trips=new Trips();
        trips.setCarNum(carNum);
        trips.setId(carInfoId);
        //获取车票详细信息
        Trips tripsInfoData=tripsDao.getTripsInfoByCarInfoIdAndId(trips);
        //判断车票是否卖光了
        Order order=new Order();
        order.setCarInfoId(carInfoId);
        order.setPersonId(customer.getId());
        order.setChangeTimes(0);
        order.setStatus(0);
        if (tripsInfoData.getTicketNum()>=1){
            orderDao.buyTicket(order);
            trips.setTicketNum(tripsInfoData.getTicketNum() - 1);
            trips.setCarNum(null);
            int i=tripsDao.updateTrips(trips);
            Map<String, Object> detailData=new HashMap<>();
            if (order.getId()>0 && i==1){
                //还有车票,购买成功
                result.setMsg("购票成功");
                result.setStateCode(200);
                detailData.put("customer", customer);
```

```java
                detailData.put("changeTimes",3 - order.getChangeTimes());
                detailData.put("order", order);
                result.setData(detailData);
            }
            return result;
        }
        else {
            //车票卖光了,购买失败
            result.setMsg("购买失败,车票已经卖光");
            result.setStateCode(400);
            result.setData(false);
            return result;
        }
    }
    @Transactional
    public Result ticketRetund(int personId , String carNum, String startTime,
        String reachTime){
        Result result=new Result();
        //票数+1
        int i=tripsDao.refundTrips(personId, carNum, startTime, reachTime);
        //将订单状态改为退票
        int j=orderDao.updateOrder1(personId, carNum, startTime, reachTime);
        if (i>0 && j>0){
            result.setData(true);
            result.setMsg("退票成功");
            result.setStateCode(200);
        }else {
            result.setData(false);
            result.setMsg("退票失败");
            result.setStateCode(400);
        }
        return result;
    }
    @Transactional
    public Result payMoney(int orderId) {
        Result result=new Result();
        if(orderDao.updateOrder(orderId)==1){
            result.setStateCode(200);
            result.setMsg("支付成功");
            result.setData(true);
        }else {
            result.setData(false);
            result.setMsg("支付失败,请重新支付");
            result.setStateCode(400);
```

```
        }
        return result;
    }
    /**
     * 保存车次
     * @param trips
     * @return
     */
    public int saveTrip(Trips trips) {
        return tripsDao.saveTrip(trips);
    }
    public int updateTripForAdmin(Trips trips){
        return tripsDao.updateTripForAdmin(trips);
    }
    public int delTrip(Integer id){
        return tripsDao.deleteTrip(id);
    }
}
```

4. Controller 层

TripsController.java 文件的源代码如下：

```
package com.code.rts.controller;
import com.code.rts.Result.Result;
import com.code.rts.entity.Trips;
import com.code.rts.entity.User;
import com.code.rts.service.TripsService;
import com.github.pagehelper.PageHelper;
import com.github.pagehelper.PageInfo;
import org.springframework.transaction.annotation.Transactional;
import org.springframework.web.bind.annotation.*;
import javax.annotation.Resource;
import java.util.HashMap;
import java.util.List;
import java.util.Map;
/**
 * 查询车票
 **/
@RestController
@CrossOrigin
public class TripsController {
    /**
     *
     */
    @Resource
    private TripsService tripsService;
```

```java
@PostMapping("/getAimtrips")
@ResponseBody
public Result getAimtrips(@RequestBody Trips trips){
    Result result=tripsService.getAimtrips(trips);
    return result;
}
@PostMapping("/getalltrips")
@ResponseBody
public Result getAlltrips(@RequestBody Trips trips){
    Result result=tripsService.getAlltrips(trips);
    return result;
}
/**
 *得到分页用户
 *@param pn
 *@return
 */
@GetMapping("/getalltripsforadmin")
public Map<String,Object> getAllTripsForAdmin(
@RequestParam(defaultValue="1",required=true,value="pn") Integer pn){
    //每页显示记录数
    Integer pageSize=5;
    //分页查询,注意顺序,startPage()方法放前面
    PageHelper.startPage(pn, pageSize);
    //获取所用用户信息
    List<Trips>allTrip=tripsService.getAllTripsForAdmin();
    //使用pageInfo包装查询后的结果,只需要将pageInfo交给页面就行。
    //  封装详细的分页信息,传入连续显示的页数
    PageInfo<Trips> pageInfo=new PageInfo(allTrip);
    Map<String,Object>modelMap=new HashMap<>();
    if (pageInfo !=null){
        modelMap.put("code", 200);
        modelMap.put("data", pageInfo);
    }else {
        modelMap.put("code", 200);
        Map<String, Object>dataMap=new HashMap<>();
        dataMap.put("message", "获取model列表失败");
        dataMap.put("entity", null);
        modelMap.put("data", dataMap);
    }
    return modelMap;
}
/**
 *保存车次信息
 *@param trips
 *@return
```

```java
 */
@Transactional
@RequestMapping(value="/saveTrip",method=RequestMethod.POST)
@ResponseBody
public Map<String, Object> saveUser(@RequestBody Trips trips){
    int i=tripsService.saveTrip(trips);
    Map<String, Object> modelMap=new HashMap<>();
    if (i==1){
        modelMap.put("code", 200);
        Map<String, Object> dataMap=new HashMap<>();
        dataMap.put("message", "success");
        dataMap.put("entity", null);
        modelMap.put("data", dataMap);
    }else {
        modelMap.put("code", 200);
        Map<String, Object> dataMap=new HashMap<>();
        dataMap.put("message", "添加车次失败");
        dataMap.put("entity", null);
        modelMap.put("data", dataMap);
    }
    return modelMap;
}

/**
 *修改用户信息
 */

@RequestMapping(value="/updateTripForAdmin",method=RequestMethod.POST)
@ResponseBody
public Map<String, Object> updateTripForAdmin(@RequestBody Trips trips){
    int i=tripsService.updateTripForAdmin(trips);
    Map<String, Object> modelMap=new HashMap<>();
    if (i==1){
        modelMap.put("code", 200);
        Map<String, Object> dataMap=new HashMap<>();
        dataMap.put("message", "success");
        dataMap.put("entity", null);
        modelMap.put("data", dataMap);
    }else {
        modelMap.put("code", 200);
        Map<String, Object> dataMap=new HashMap<>();
        dataMap.put("message", "更新车次信息失败");
        dataMap.put("entity", null);
        modelMap.put("data", dataMap);
    }
    return modelMap;
```

```java
        }

        /**
         *根据id删除车次
         */
        @Transactional
        @RequestMapping(value="/deleteTrip/{id}",method=RequestMethod.DELETE)
        @ResponseBody
        public Map<String, Object> deleteTrip(@PathVariable("id") Integer id){
            Map<String, Object> modelMap=new HashMap<>();

            try {
                int i=tripsService.delTrip(id);
                if (i==1){
                    modelMap.put("code", 200);
                    Map<String, Object> dataMap=new HashMap<>();
                    dataMap.put("message", "success");
                    dataMap.put("entity", null);
                    modelMap.put("data", dataMap);
                }else {
                    modelMap.put("code", 200);
                    Map<String, Object> dataMap=new HashMap<>();
                    dataMap.put("message", "删除失败");
                    dataMap.put("entity", null);
                    modelMap.put("data", dataMap);
                }
            }catch (Exception e){
                modelMap.put("code", 500);
                Map<String, Object> dataMap=new HashMap<>();
                dataMap.put("message", "删除失败");
                dataMap.put("entity", null);
                modelMap.put("data", dataMap);
            }
            return modelMap;
        }
    }
```

5. View 层(略)

7.4.2 订单管理

1. 持久化层

(1) Order.java 文件的源代码如下：

```java
package com.code.rts.entity;
public class Order {
```

```java
    private int id;
    private int carInfoId;
    private int personId;
    private int changeTimes;
    //值为 0、1 或者 2;0 表示预定未付款,1 表示已经支付,2 表示退票
    private int status;
    private String stautsMsg;
    //省略 getter、setter 方法
}
```

(2) OrderReturn.java 文件的源代码如下：

```java
package com.code.rts.entity;
public class OrderReturn {
    private Integer id;
    private String orginLocation;
    private String destinationLocation;
    private String startTime;
    private String reachTime;
    private String carNum;
    private int ticketPrice;
    private int ticketNum;
    private String trueName;
    private String idCardNum;
    private String phoneNum;
    private String status;
    //省略 getter、setter 方法
}
```

2. DAO 层

(1) OrderDao.java 文件的源代码如下：

```java
package com.code.rts.dao;
import com.code.rts.entity.*;
import com.github.pagehelper.Page;
import org.apache.ibatis.annotations.*;
import org.aspectj.weaver.ast.Or;
import java.util.List;
public interface OrderDao {
    /**
     *查询所有订单
     *@return
     */
    Page<OrderReturn> getAllOrders();
    /**
     *删除订单
```

```java
 * @param id
 * @return
 */
Integer deleteOrder(Integer id);
/**
 *插入订单信息
 * @param order
 * @return
 */
@Insert("INSERT INTO 'order' (car_info_id, person_id, change_times, status)" +
    " VALUES (#{carInfoId}, #{personId}, #{changeTimes}, #{status})\n" + " ")
@Options(useGeneratedKeys=true, keyProperty="id", keyColumn="id")
void buyTicket(Order order);
/**
 *下订单(状态:未支付)
 * @param order
 * @return
 */
@Insert("INSERT INTO 'order' (car_info_id, person_id, change_times, status)
    VALUES (#{carInfoId}, #{personId}, #{changeTimes}, #{status})")
@Options(useGeneratedKeys=true, keyProperty="id", keyColumn="id")
int saveOrderPaying(Order order);
/**
 * @param order
 * @return
 */
@Update("update'order' set status=#{status} where id=#{id}")
int updateOrderStatus(Order order);
/**
 *退票,将 status 改为 3
 * @param orderId
 * @return
 */
@Update("update 'order' set status=3 where id=#{orderId}")
int updateOrder(int orderId);
/**
 *支付后,将 status 改成 2
 * @param orderId
 * @return
 */
@Update("update'order' set status=2 where id=#{orderId}")
int saveOrderPayed(int orderId);
/**
 *通过用户查询订单(前台使用)
```

```java
 * @param userName
 * @return
 */
Page<OrderReturn> getOrder(String userName);
@Update("update 'order' set status=2 where person_id=#{personId} and car_info_id=
    (select id from trips where car_num=#{carNum}" + " and start_time=
    #{startTime} and reach_time=#{reachTime})")
int updateOrder1(@Param("personId") int personId, @Param("carNum")
    String carNum,@Param("startTime") String startTime,
    @Param("reachTime") String reachTime);
/**
 *查询目标订单
 * @param orderId
 * @return
 **/
@Select("select * from 'order' where id=#{orderId}")
Order getAimOrder(int orderId);
/**
 *查询所有人的订单信息
 */
@Select("select * from 'person' where 'id'=#{personId}")
User getUserinfo(int id);
/**
 *改签订单信息变更
 * @return
 */
@Update("update 'order' set car_info_id=#{tripsId},change_times=
    change_times+1,status=1 where id=#{orderId}")
int changeOrder(@Param("orderId") int orderId, @Param("tripsId") int tripsId);
}
```

(2) OrderDao.xml 文件的源代码如下：

```xml
<?xml version="1.0" encoding="UTF-8"?>
<!DOCTYPE mapper
        PUBLIC "-//mybatis.org//DTD Mapper 3.0//EN"
        "http://mybatis.org/dtd/mybatis-3-mapper.dtd">
<!-- 对应接口,写入SQL语句 -->
<mapper namespace="com.code.rts.dao.OrderDao">
    <select id="getAllOrders" resultType="com.code.rts.entity.OrderReturn"
        parameterType="com.github.pagehelper.Page">
        SELECT o.id, t.orgin_location, t.destination_location, t.start_time, t.reach_time,
            t.car_num, t.ticket_price, u.true_name, u.id_card_num, u.phone_num, o.status
        FROM 'order' o, 'user' u, 'trips' t
        WHERE o.person_id=u.id AND o.car_info_id=t.id
```

```xml
        </select>
        <select id="getOrder" resultType="com.code.rts.entity.OrderReturn"
            parameterType="com.github.pagehelper.Page">
            select * from 'order','user','trips' where 'user'.id='order'.person_id and
                'order'.car_info_id='trips'.id and user.username=#{userName}
        </select>
        <!-- 根据id删除订单 -->
        <delete id="deleteOrder" parameterType="Integer">
            DELETE FROM 'order' WHERE id=#{id}
        </delete>
</mapper>
```

3. Service层

OrderService.java文件的源代码如下：

```java
package com.code.rts.service;
import com.code.rts.Result.Result;
import com.code.rts.dao.*;
import com.code.rts.entity.*;
import com.github.pagehelper.Page;
import org.aspectj.weaver.ast.Or;
import org.springframework.stereotype.Service;
import javax.annotation.Resource;
import java.util.ArrayList;
import java.util.List;
@Service
public class OrderService {
    @Resource
    private OrderDao orderDao;
    @Resource
    private TripsDao tripsDao;
    /**
     * 获得订单状态
     * @return
     */
    public Page<OrderReturn> getAllOrders(){
        return orderDao.getAllOrders();
    }
    /**
     * 获得订单
     * @param username
     * @return
     */
    public Page<OrderReturn> getOrder(String username){
        return orderDao.getOrder(username);
```

```java
    }
    public Result changeOrder(int orderId, int tripsId) {
        Result result=null;
        Order order=orderDao.getAimOrder(orderId);
        Trips trips=tripsDao.gettrips(tripsId);
        if(trips.getTicketNum()>0){
            tripsDao.changeOldtrips(order.getCarInfoId());
            tripsDao.changeNewtrips(tripsId);
            orderDao.changeOrder(orderId,tripsId);
            result.setStateCode(200);
            result.setMsg("change order succeed");
        }
        else{
            result.setStateCode(404);
            result.setMsg("change order failed");
        }
        return result;
    }
    public int updateOrderStatus(Order order){
        return orderDao.updateOrderStatus(order);
    }
    public Integer deleteOrder(Integer id){
        return orderDao.deleteOrder(id);
    }
    public int saveOrderPaying(Order order){
        return orderDao.saveOrderPaying(order);
    }
    public int saveOrderPayed(int orderId){
        return orderDao.saveOrderPayed(orderId);
    }
}
```

4. Controller 层

OrderController.java 文件的源代码如下：

```java
package com.code.rts.controller;
import com.alibaba.fastjson.JSONObject;
import com.code.rts.Result.Result;
import com.code.rts.entity.Order;
import com.code.rts.entity.OrderReturn;
import com.code.rts.entity.User;
import com.code.rts.service.OrderService;
import com.github.pagehelper.Page;
import com.github.pagehelper.PageHelper;
import com.github.pagehelper.PageInfo;
```

```java
import org.springframework.transaction.annotation.Transactional;
import org.springframework.web.bind.annotation.*;
import javax.annotation.Resource;
import java.util.HashMap;
import java.util.List;
import java.util.Map;
@RestController
@CrossOrigin
public class OrderController {
    @Resource
    private OrderService orderService;
    /**
     * 获得分页用户
     * @param pn
     * @return
     */
    @GetMapping("/getallorders")
    public Map<String, Object> getAllOrders(@RequestParam(
        defaultValue="1",required=true,value="pn") Integer pn){
        //每页显示记录数
        Integer pageSize=5;
        //分页查询,注意顺序,startPage()方法放前面
        PageHelper.startPage(pn, pageSize);
        //获取所用用户信息
        List<OrderReturn> allOrder=orderService.getAllOrders();
        //使用pageInfo包装查询后的结果,只需要将pageInfo交给页面就行。
        //  封装详细的分页信息,传入连续显示的页数
        PageInfo<OrderReturn> pageInfo=new PageInfo(allOrder);
        Map<String, Object> modelMap=new HashMap<>();
        if (pageInfo!=null){
            modelMap.put("code", 200);
            modelMap.put("data", pageInfo);
        }else {
            modelMap.put("code", 200);
            Map<String, Object> dataMap=new HashMap<>();
            dataMap.put("message", "获取订单列表失败");
            dataMap.put("entity", null);
            modelMap.put("data", dataMap);
        }
        return modelMap;
    }
    /**
     * 通过用户名获得分页订单
     * @param pn
```

```java
 * @return
 */
@GetMapping("/getorder")
public Map<String, Object> getOrder(@RequestParam(defaultValue=
    "1",required=true,value="pn")String username, Integer pn){
    //每页显示记录数
    Integer pageSize=10;
    //分页查询,注意顺序,startPage()方法放前面
    PageHelper.startPage(pn, pageSize);
    //获取所用用户信息
    Page<OrderReturn> allOrder=orderService.getOrder(username);
    //使用pageInfo包装查询后的结果,只需要将pageInfo交给页面就行。
    封装详细的分页信息,传入连续显示的页数
    PageInfo<OrderReturn> pageInfo=new PageInfo(allOrder);

    Map<String, Object> modelMap=new HashMap<>();
    if (pageInfo !=null){
        modelMap.put("code", 200);
        modelMap.put("data", pageInfo);
    }else {
        modelMap.put("code", 200);
        Map<String, Object> dataMap=new HashMap<>();
        dataMap.put("message", "获取订单列表失败");
        dataMap.put("entity", null);
        modelMap.put("data", dataMap);
    }
    return modelMap;
}
/**
 * change order
 * @return
 **/
@PostMapping("/changeorder")
public Result changeOrder(@RequestBody JSONObject jsonObject){
    int orderid=jsonObject.getInteger("orderId");
    int tripsid=jsonObject.getInteger("tripsId");
    Result result=orderService.changeOrder(orderid,tripsid);
    return result;
}
/**
 *修改订单信息
 *@Param: orderID 和 status
 */
@RequestMapping(value="/updateorder",method=RequestMethod.POST)
```

```java
@ResponseBody
public Map<String, Object> updateUser(@RequestBody Order order){
    int i=orderService.updateOrderStatus(order);
    Map<String, Object> modelMap=new HashMap<>();
    if (i==1){
        modelMap.put("code", 200);
        Map<String, Object> dataMap=new HashMap<>();
        dataMap.put("message", "success");
        dataMap.put("entity", null);
        modelMap.put("data", dataMap);
    }else {
        modelMap.put("code", 200);
        Map<String, Object> dataMap=new HashMap<>();
        dataMap.put("message", "获取model列表失败");
        dataMap.put("entity", null);
        modelMap.put("data", dataMap);
    }
    return modelMap;
}
/**
 *根据id删除订单
 */
@Transactional
@RequestMapping(value="/deleteOrder/{id}",method=RequestMethod.DELETE)
@ResponseBody
public Map<String, Object> deleteOrder(@PathVariable("id") Integer id){
    Map<String, Object> modelMap=new HashMap<>();
    //删除用户
    try {
        int i=orderService.deleteOrder(id);
        if (i==1){
            modelMap.put("code", 200);
            Map<String, Object> dataMap=new HashMap<>();
            dataMap.put("message", "success");
            dataMap.put("entity", null);
            modelMap.put("data", dataMap);
        }else {
            modelMap.put("code", 200);
            Map<String, Object> dataMap=new HashMap<>();
            dataMap.put("message", "删除失败");
            dataMap.put("entity", null);
            modelMap.put("data", dataMap);
        }
    }catch (Exception e){
```

```java
            modelMap.put("code", 500);
            Map<String, Object> dataMap=new HashMap<>();
            dataMap.put("message", e.fillInStackTrace());
            dataMap.put("entity", null);
            modelMap.put("data", dataMap);
    }
    return modelMap;
}
/**
 *下订单(状态:未支付)
 */
@Transactional
@RequestMapping(value="/saveOrderPaying",method=RequestMethod.POST)
@ResponseBody
public Map<String, Object> deleteOrder(@RequestBody JSONObject jsonObject){
    int carInfoId=jsonObject.getInteger("car_info_id");
    int personId=jsonObject.getInteger("person_id");
    Order order=new Order();
    order.setCarInfoId(carInfoId);
    order.setPersonId(personId);
    order.setChangeTimes(0);
    order.setStatus(1);
    Map<String, Object> modelMap=new HashMap<>();
    try {
            //此处应该获取产生订单后的id号
            int i=orderService.saveOrderPaying(order);
            int orderId=order.getId();
            if (i==1){
                modelMap.put("code", 200);
                Map<String, Object> dataMap=new HashMap<>();
                dataMap.put("message", "success");
                dataMap.put("entity", orderId);
                modelMap.put("data", dataMap);
            }else {
                modelMap.put("code", 200);
                Map<String, Object> dataMap=new HashMap<>();
                dataMap.put("message", "购买失败");
                dataMap.put("entity", null);
                modelMap.put("data", dataMap);
            }
    }catch (Exception e){
            modelMap.put("code", 500);
            Map<String, Object> dataMap=new HashMap<>();
            dataMap.put("message", e.fillInStackTrace());
```

```java
                dataMap.put("entity", null);
                modelMap.put("data", dataMap);
            }
            return modelMap;
        }
        @Transactional
        @RequestMapping(value="/saveOrderPayed/{id}",method=RequestMethod.GET)
        @ResponseBody
        public Map<String, Object> saveOrderPayed(@PathVariable("id")Integer id){
            Map<String, Object> modelMap=new HashMap<>();
            try {
                int i=orderService.saveOrderPayed(id);
                if (i==1){
                    modelMap.put("code", 200);
                    Map<String, Object> dataMap=new HashMap<>();
                    dataMap.put("message", "success");
                    dataMap.put("entity", null);
                    modelMap.put("data", dataMap);
                }else {
                    modelMap.put("code", 200);
                    Map<String, Object> dataMap=new HashMap<>();
                    dataMap.put("message", "操作失败");
                    dataMap.put("entity", null);
                    modelMap.put("data", dataMap);
                }
            }catch (Exception e){
                modelMap.put("code", 500);
                Map<String, Object> dataMap=new HashMap<>();
                dataMap.put("message", e.fillInStackTrace());
                dataMap.put("entity", null);
                modelMap.put("data", dataMap);
            }
            return modelMap;
        }
    }
```

5. View 层(略)

7.5 项目搭建

火车订票系统使用 Idea 工具和 MySQL 数据库开发，系统项目结构如图 7-9、图 7-10 所示。

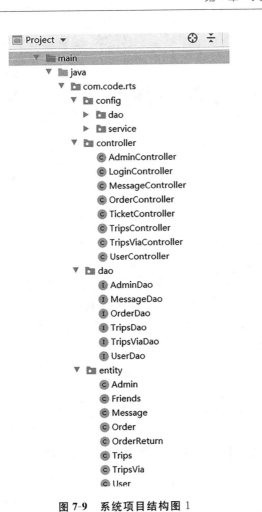

图 7-9 系统项目结构图 1

图 7-10 系统项目结构图 2

第8章 员工管理系统

8.1 需求分析

8.1.1 系统概述

随着计算机技术的飞速发展,计算机运用逐渐深入各行各业,给人的生活、工作带来了巨大变化。在企业中,采用计算机管理员工的方式已经逐步代替了采用手工记录管理员工的方式,因此,各种企业人事管理软件也随之诞生。企业的员工管理正朝着高效率、低成本、稳定可靠的方向发展,因此,企业需要一个功能完善、性能稳定的员工管理系统。企业员工管理工作是目前企业各项工作中的重点工作之一,如何能科学有效地开展好管理工作,是当前企业普遍关心的问题。因此,实现企业员工的信息化管理有着十分重要的意义。

8.1.2 功能需求描述

员工管理是整个人事管理流程最重要的环节,包含员工人事档案多种维度管理、员工信息录入、员工花名册的导入和导出过程。员工管理系统可以真正实现员工信息无纸化管理。同时,可以使用员工管理系统创建和编辑部门,也可以在员工花名册导入时自动读取员工部门层级信息。

员工管理系统主要包含以下几个方面。

(1) 用户管理:注册、登录、注销、用户列表分页展示。
(2) 角色管理:管理用户对应的角色及角色对应的资源。
(3) 部门管理:根据部门的上下级关系,完成对部门的树形展示。
(4) 员工管理:管理部门所属员工。
(5) 日志管理:记录用户登录日志,记录用户操作日志。

8.2 总体设计

8.2.1 系统总体功能结构

通过员工管理系统分析,其功能模块图如图8-1所示。

8.2.2 总体架构

员工管理系统采用SpringBoot开发,其简化了开发过程、配置过程、部署过程和监控过程。项目架构如图8-2所示。

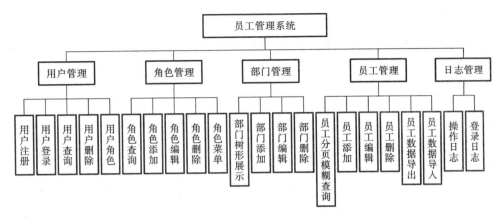

图 8-1　员工管理系统功能模块图

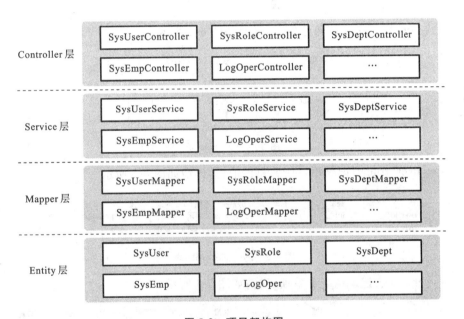

图 8-2　项目架构图

SysUserController、SysUserService、SysUserMapper、SysUser 分别实现用户管理模块的控制层、数据服务接口层、数据接口访问层和实体层的功能；SysDeptController、SysDeptService、SysDeptMapper、SysDept 分别实现部门管理模块的控制层、数据服务接口层、数据接口访问层和实体层的功能；SysEmpController、SysEmpService、SysEmpMapper、SysEmp 分别实现员工管理模块的控制层、数据服务接口层、数据接口访问层和实体层的功能；SysRoleController、SysRoleService、SysRoleMapper、SysRole 分别实现角色管理模块的控制层、数据服务接口层、数据接口访问层和实体层的功能；LogOperController、LogOperService、LogOperMapper、LogOper 分别实现日志操作模块的控制层、数据服务接口层、数据接口访问层和实体层的功能。

8.2.3 数据库设计

(1) sys_user 表记录用户的个人信息,详情如图 8-3 所示。

名	类型	长度	小数点	不是 null	虚拟	键	注释
user_id	int			☑	☐	🔑1	主键id
user_name	varchar	25		☐	☐		用户名
nick_name	varchar	25		☐	☐		用户昵称
password	varchar	100		☐	☐		用户密码
email	varchar	25		☐	☐		邮箱
phone	varchar	25		☐	☐		手机号码
avatar	varchar	255		☐	☐		头像
create_time	datetime			☐	☐		创建时间
update_time	datetime			☐	☐		更新时间
enabled	bit	1		☐	☐		是否可用
del_flag	bit	1		☐	☐		是否删除

图 8-3 sys_user 表

(2) sys_role 表记录系统角色信息,详情如图 8-4 所示。

名	类型	长度	小数点	不是 null	虚拟	键	注释
role_id	int			☑	☐	🔑1	主键id
role_name	varchar	255		☐	☐		角色名称
role_key	varchar	255		☐	☐		角色权限标识
del_flag	bit	1		☐	☐		是否删除

图 8-4 sys_role 表

(3) sys_user_role 表记录用户所属角色信息,详情如图 8-5 所示。

名	类型	长度	小数点	不是 null	虚拟	键	注释
user_id	int			☑	☐	🔑1	用户ID
role_id	int			☑	☐	🔑2	角色ID

图 8-5 sys_user_role 表

(4) sys_menu 表记录系统菜单项,详情如图 8-6 所示。

名	类型	长度	小数点	不是 null	虚拟	键	注释
menu_id	int			☑	☐	🔑1	主键id
menu_name	varchar	25		☐	☐		菜单名称
parent_id	int			☐	☐		父级菜单id
url	varchar	100		☐	☐		访问url
path	varchar	100		☐	☐		路由path
component	varchar	25		☐	☐		组件
icon	varchar	100		☐	☐		图标
enabled	bit	1		☐	☐		是否可用

图 8-6 sys_menu 表

(5) sys_role_menu 表记录角色与菜单项的关联,详情如图 8-7 所示。

(6) sys_dept 表记录部门信息,详情如图 8-8 所示。

(7) sys_emp 表记录员工信息,详情如图 8-9 所示。

名	类型	长度	小数点	不是 null	虚拟	键	注释
role_id	int			☑	☐	🔑1	角色ID
menu_id	int			☑	☐	🔑2	菜单ID

图 8-7 sys_role_menu 表

名	类型	长度	小数点	不是 null	虚拟	键	注释
dept_id	int			☑	☐	🔑1	部门id
parent_id	int			☐	☐		父级部门id
ancestors	varchar	100		☐	☐		祖宗节点
dept_name	varchar	255		☐	☐		部门名称
create_time	datetime			☐	☐		创建时间
update_time	datetime			☐	☐		更新时间

图 8-8 sys_dept 表

名	类型	长度	小数点	不是 null	虚拟	键	注释
emp_id	int			☑	☐	🔑1	员工id
dept_id	int			☐	☐		部门id
emp_name	varchar	20		☐	☐		员工名称
gender	varchar	1		☐	☐		性别
birthday	date			☐	☐		出生日期
native_place	varchar	25		☐	☐		籍贯
nation	varchar	25		☐	☐		民族
id_card	varchar	18		☐	☐		身份证号码
phone	varchar	11		☐	☐		手机号码
email	varchar	20		☐	☐		邮箱
degree	varchar	25		☐	☐		学位
position	varchar	25		☐	☐		职位
work_time	date			☐	☐		入职时间
create_time	datetime			☐	☐		创建时间
update_time	datetime			☐	☐		修改时间

图 8-9 sys_emp 表

（8）ser_email 表记录邮件信息，详情如图 8-10 所示。

名	类型	长度	小数点	不是 null	虚拟	键	注释
email_id	int			☑	☐	🔑1	主键id
addressee	varchar	255		☐	☐		收件人
email_subject	varchar	255		☐	☐		邮件主题
email_text	text			☐	☐		邮件内容
sent_date	datetime			☐	☐		发送时间

图 8-10 ser_email 表

（9）ser_notice 表记录消息，详情如图 8-11 所示。

（10）log_oper 表记录系统操作信息，详情如图 8-12 所示。

（11）log_login 表记录登录的日志信息，详情如图 8-13 所示。

名	类型	长度	小数点	不是 null	虚拟	键	注释
notice_id	int			☑	☐	🔑1	主键id
notice_title	varchar	100		☐	☐		标题
notice_type	enum			☐	☐		类型
notice_content	text			☐	☐		内容
create_time	datetime			☐	☐		创建时间
update_time	datetime			☐	☐		更新时间

图 8-11　ser_notice 表

名	类型	长度	小数点	不是 null	虚拟	键	注释
oper_id	int			☑	☐	🔑1	主键id
oper_module	varchar	25		☐	☐		操作模块
oper_content	varchar	25		☐	☐		操作内容
user_name	varchar	25		☐	☐		操作用户名
oper_ip	varchar	25		☐	☐		ip
oper_location	varchar	25		☐	☐		地址
oper_status	varchar	25		☐	☐		操作状态
oper_time	datetime			☐	☐		操作时间

图 8-12　log_oper 表

名	类型	长度	小数点	不是 null	虚拟	键	注释
login_id	int			☑	☐	🔑1	主键ID
user_name	varchar	255		☐	☐		登录用户名
ip	varchar	255		☐	☐		登录IP
location	varchar	255		☐	☐		登录地址
browser	varchar	25		☐	☐		浏览器
os	varchar	25		☐	☐		操作系统
login_time	datetime			☐	☐		登录时间

图 8-13　log_login 表

8.3　详细设计

8.3.1　用户管理

1. 用户注册

后台生成 4 位随机字符串，存储到 redis 中，并且设置 60 s 的过期时间。通过用户输入验证码进行校验，如果 redis 获取验证码为空，则说明验证码过期。

2. 用户登录

使用 Spring Security 进行用户登录，配置登录请求接口/doLogin，并传入用户名和密码，再实现 UserDetailService 接口配置数据源为 MySQL。只需要传入用户名对应的用户对象，密码校验由 Spring Security 自动完成。

3. 用户查询

使用 MyBatis 查询 sys_user 表的所有记录,并映射到 List⟨SysUser⟩ 集合中,通过控制 (Controller)层提供接口/user/list,将 List⟨SysUser⟩ 集合序列化为 json 对象,再返回给前端调用,前端获取 json 数据并渲染到页面上。

4. 用户编辑

前端对用户进行编辑并提交表单后,服务端接收编辑后的用户对象,由 Controller 层传到 Mapper 层并进行 SQL 编辑,根据用户的主键 id 修改数据库。

5. 用户删除(逻辑删除)

前端提交要删除用户的主键 ID,服务端接收编辑后的用户对象,由 Controller 层传到 Mapper 层并进行 SQL 编辑,根据用户的主键 id 删除数据库。

6. 用户角色(多对多)

根据 sys_user、sys_role、sys_user_role 三张表查询用户对应的角色集合,将用户对应的角色集合响应到前端并进行渲染。

用户添加页面如图 8-14 所示,用户添加成功后的页面如图 8-15 所示。

图 8-14 用户添加页面

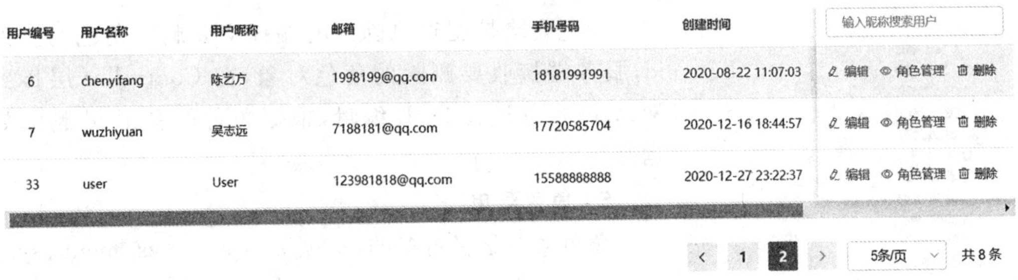

图 8-15 用户添加成功后的页面

8.3.2 角色管理

1. 角色查询

使用 MyBatis 查询 sys_role 表的所有记录,映射到 List⟨SysRole⟩ 集合中,通过 Controller 层提供接口/role/list,将 List⟨SysRole⟩ 集合序列化为 json 数据,再返回给前端调用,前端获取 json 数据并渲染到页面上。

2. 角色添加

前端提交表单后,服务端接收添加后的角色对象,由 Controller 层传到 Mapper 层并执行 SQL 命令,再将命令添加到数据库。

3. 角色编辑

前端对角色进行编辑并提交后,服务端接收编辑后的角色对象,由 Controller 层传到 Mapper 层并进行 SQL 编辑,根据角色的主键 id 修改数据库。用户选择编辑角色如图 8-16 所示。用户绑定角色后的页面如图 8-17 所示。

图 8-16 用户选择编辑角色

图 8-17 用户绑定角色后的页面

4. 角色删除

角色删除是逻辑删除。前端提交要删除角色的主键 id,服务端接收要删除的角色对象,由 Controller 层传到 Mapper 层并进行 SQL 编辑,根据角色的主键 id 删除数据库。

5. 角色菜单

角色菜单是多对多的,根据 sys_role、sys_menu、sys_role_menu 三张表查询角色对应的菜单集合,将角色对应的菜单集合响应到前端并进行渲染。给不同的角色绑定菜单项如图 8-18 所示。

图 8-18 给不同的角色绑定菜单项

8.3.3 部门管理

1. 部门树形展示

根据部门的主键 id 和部门的 parent_id(父级部门主键 id),使用 MyBatis 的递归查询 sys_dept 表的所有记录,映射到 List<SysDept>集合中,通过 Controller 层提供接口/dept/list,将 List<SysDept>集合序列化为 json 数据,再返回给前端调用,前端获取 json 数据并渲染到页面上。部门查询页面如图 8-19 所示。

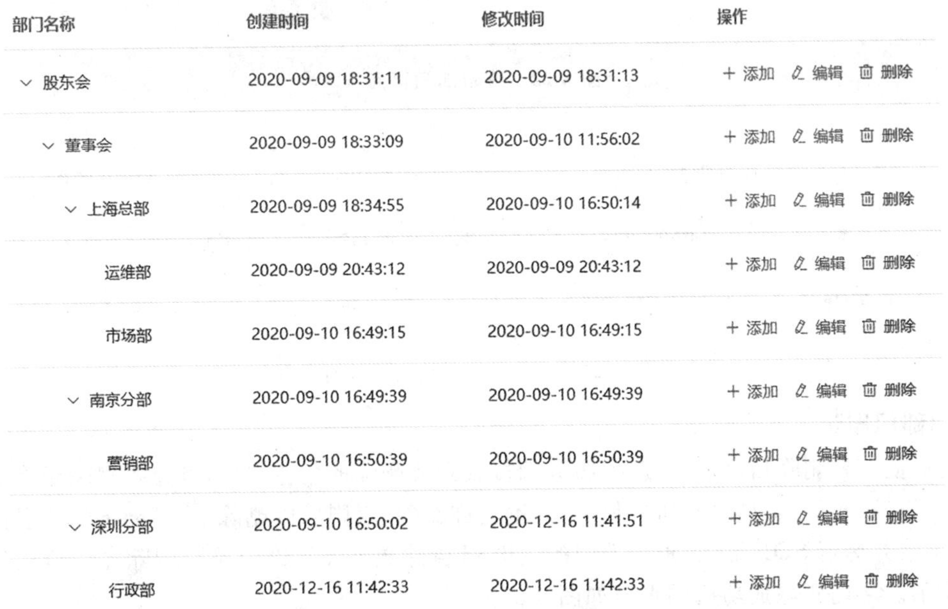

图 8-19 部门查询页面

2. 部门添加

前端提交表单后,服务端接收到添加后的部门对象,由 Controller 层传到 Mapper 层并执行 SQL 命令,再将命令添加到数据库。添加部门页面如图 8-20 所示。添加部门成功后的页面如图 8-21 所示。

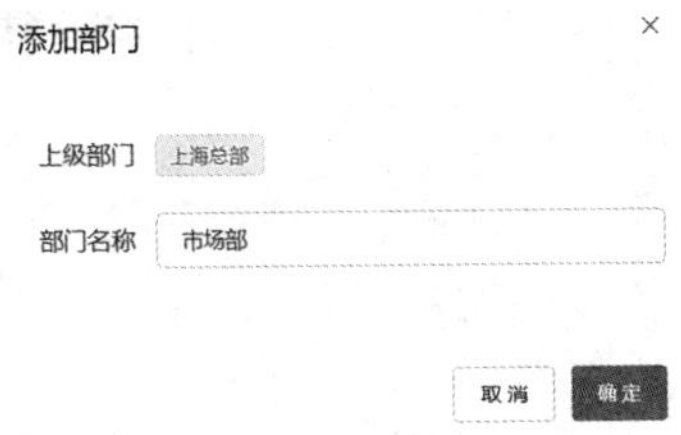

图 8-20 添加部门页面

3. 部门编辑

前端对部门进行编辑并提交后,服务端接收到编辑后

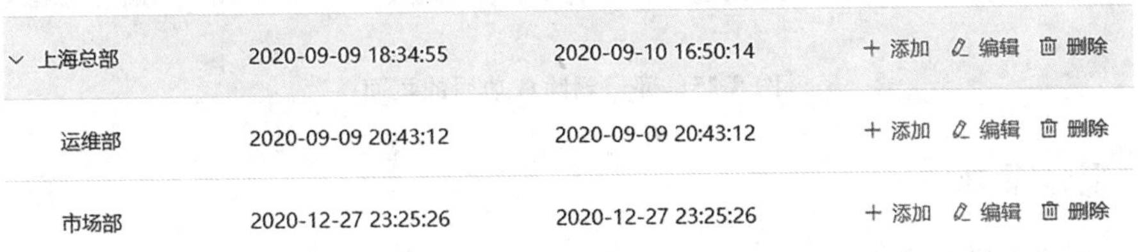

图 8-21 部门添加成功后的页面

的部门对象,由 Controller 层传到 Mapper 层并运行 SQL 语句,根据部门的主键 id 修改数据库。编辑部门页面如图 8-22 所示。编辑部门成功后的页面如图 8-23 所示。

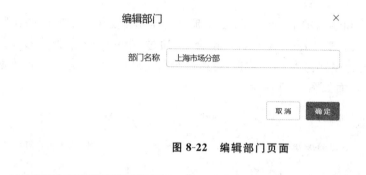

图 8-22 编辑部门页面

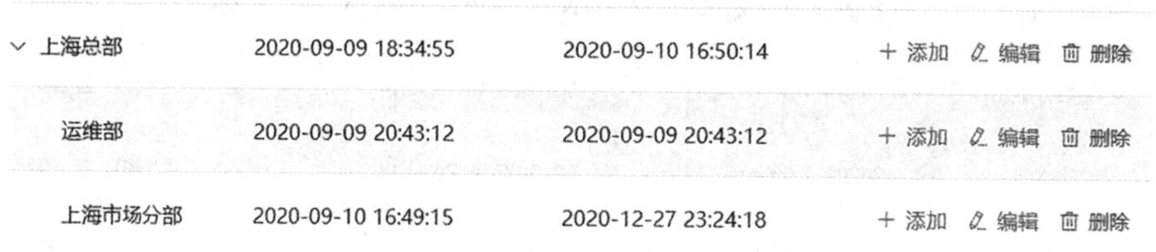

图 8-23 编辑部门成功后的页面

4. 部门删除

前端提交要删除角色的主键 id,服务端接收到要删除的角色对象,首先要判断删除的部门下是否有所属员工,如果有所属员工,则不允许删除,否则可以删除,由 Controller 层传到 Mapper 层并运行 SQL 语句,根据部门的主键 id 删除数据库。部门是否删除提示信息如图 8-24 所示。部门删除成功后的页面如图 8-25 所示。

图 8-24 部门是否删除提示信息

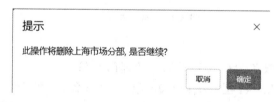

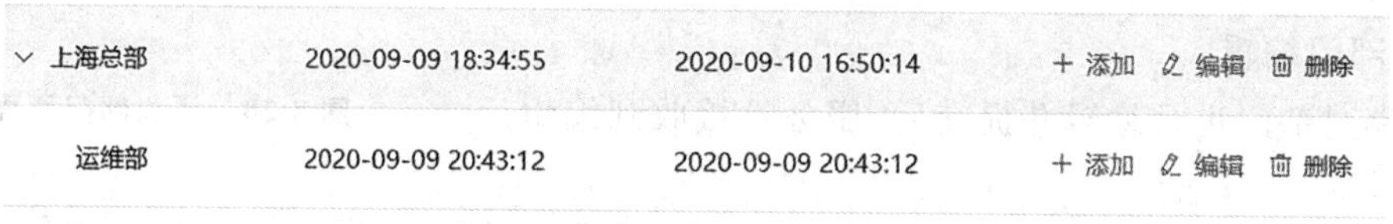

图 8-25 部门删除成功后的页面

8.3.4 员工管理

1. 员工分页模糊查询

接收前端的模糊查询条件,使用 MyBatis 动态查询所有员工,控制层将 json 格式数据响应给前端,前端使用 Element UI 的分页组件进行分页展示。员工查询页面如图 8-26 所示。

图 8-26 员工查询页面

2. 员工添加

前端提交表单后,服务端接收到添加后的角色对象,由 Controller 层传到 Mapper 层并执行 SQL 命令,再添加到数据库。员工添加页面如图 8-27 所示。员工添加后的结果如图 8-28 所示。

图 8-27 员工添加页面

图 8-28 员工添加后的结果

3. 员工编辑

前端对角色进行编辑并提交表单后,服务端接收到编辑后的角色对象,由 Controller 层传到 Mapper 层并进行 SQL 编辑,根据角色的主键 id 修改数据库。员工编辑页面如图 8-29 所示。员工编辑成功后的页面如图 8-30 所示。

4. 员工删除

前端提交要删除角色的主键 id,服务端接收到要删除的角色对象,由 Controller 层传到

图 8-29 员工编辑页面

图 8-30 员工编辑成功后的页面

Mapper 层并进行 SQL 编辑，根据角色的主键 id 删除数据库。员工是否删除提示信息如图 8-31 所示。员工删除成功后的页面如图 8-32 所示。

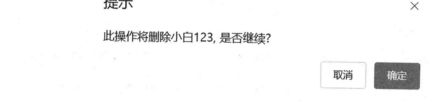

图 8-31 员工是否删除提示信息

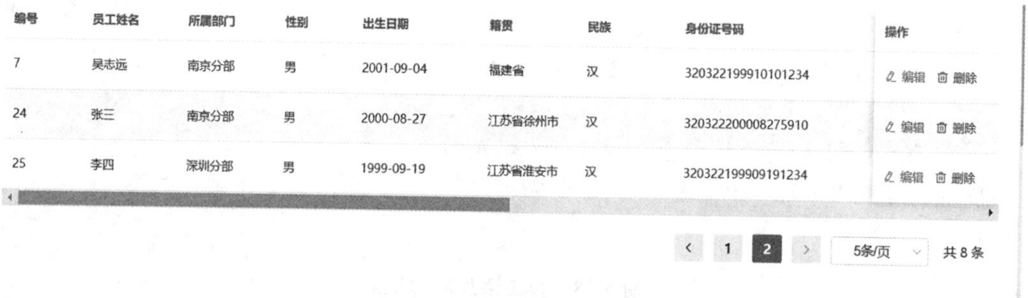

图 8-32 员工删除成功后的页面

5. 员工数据导出

使用 Easy POI 对实体类对象 SysEmp 进行注解式配置，@ExcelTarget 和 @Excel 为配置属性，使用 ExcelExportUtil 导出 Excel 表格。员工数据导出到 Excel，如图 8-33 所示。

6. 员工数据导入

从前端上传 Excel 文件到后端，后端接收参数为 MultipartFile，使用 ExcelImportUtil

员工编号	员工姓名	所属部门	性别	出生日期	籍贯	民族	身份证号码	手机号码	邮箱	学位	职位	任职时间	创建时间
1	张昊	股东会	男	2000-08-27	江苏省徐州市	汉	320322200008275910	18114970827	785896760@qq.com	本科	Java工程师	2020-09-03	2020-09-10 13:28:43
2	孙岩123	董事会	男	1999-09-19	江苏省淮安市	汉	320322199909191234	19852105211	1572837278@qq.com	本科	C++工程师	2020-09-04	2020-09-10 13:37:18
3	王拓宾	董事会	女	1999-04-28	山西省运城市	汉	142726199904283011	15985968023	1093824058@qq.com	本科	算法工程师	2020-09-05	2020-09-10 13:39:53
4	徐致为	南京分部	男	2000-06-26	湖北省武汉市	汉	420302200006251518	13190202911	2217332694@qq.com	本科	前端工程师	2020-09-06	2020-09-10 13:41:44
6	陈艺方	深圳分部	男	1999-11-25	山东省日照市	汉	320322199910101234	15863371897	1189191010@qq.com	本科	运维工程师	2020-08-31	2020-09-12 14:26:53
7	吴志远	南京分部	男	2001-09-04	福建省	汉	320322199910101234	18114970827	17281882@qq.com	本科	算法工程师	2020-09-07	2020-09-12 15:05:41
24	张三	南京分部	男	2000-08-27	江苏省徐州市	汉	320322200008275910	18114970827	785896760@qq.com	本科	Java工程师	2020-09-03	2020-09-10 13:28:43
25	李四	深圳分部	男	1999-09-19	江苏省淮安市	汉	320322199909191234	19852105211	1572837278@qq.com	本科	C++工程师	2020-09-04	2020-09-10 13:37:18

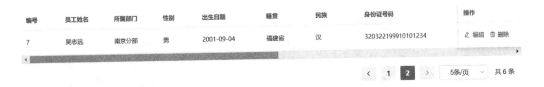

图 8-33　员工数据导出到 Excel

的 importExcel 方法解析文件为员工列表，将员工列表添加到数据库。员工数据从 Excel 文档导入系统，分别如图 8-34 至图 8-36 所示。

编号	员工姓名	所属部门	性别	出生日期	籍贯	民族	身份证号码	操作
7	吴志远	南京分部	男	2001-09-04	福建省	汉	320322199910101234	编辑 删除

‹ 1 2 › 5条/页 共 6 条

图 8-34　员工数据导入 Excel 之前的数据

员工编号	员工姓名	所属部门	性别	出生日期	籍贯	民族	身份证号码	手机号码	邮箱	学位	职位	任职时间	创建时间
24	张三	南京分部	男	2000-08-27	江苏省徐州市	汉	320322200008275910	18114970827	785896760@qq.com	本科	Java工程师	2020-09-03	2020-09-10 13:28:43
25	李四	深圳分部	男	1999-09-19	江苏省淮安市	汉	320322199909191234	19852105211	1572837278@qq.com	本科	C++工程师	2020-09-04	2020-09-10 13:37:18

图 8-35　Excel 数据

编号	员工姓名	所属部门	性别	出生日期	籍贯	民族	身份证号码	操作
7	吴志远	南京分部	男	2001-09-04	福建省	汉	320322199910101234	编辑 删除
27	张三	南京分部	男	2000-08-27	江苏省徐州市	汉	320322200008275910	编辑 删除
28	李四	深圳分部	男	1999-09-19	江苏省淮安市	汉	320322199909191234	编辑 删除

‹ 1 2 › 5条/页 共 8 条

图 8-36　员工数据导入 Excel 之后的数据

8.3.5　日志管理

1. 操作日志

配置 Log 注解和 LogAspect 切面类。@PointCut 为配置切点，@AfterReturning 为配置请求后的执行方法，@AfterThrowing 为配置异常处理方法。查看操作日志如图 8-37 所示。

2. 登录日志

用户登录成功后，调用记录登录日志的方法，使用 UserAgentUtils 类解析出登录的 IP 地址及登录地点，封装成 SysLogin 对象添加到数据库中。查看登录日志如图 8-38 所示。

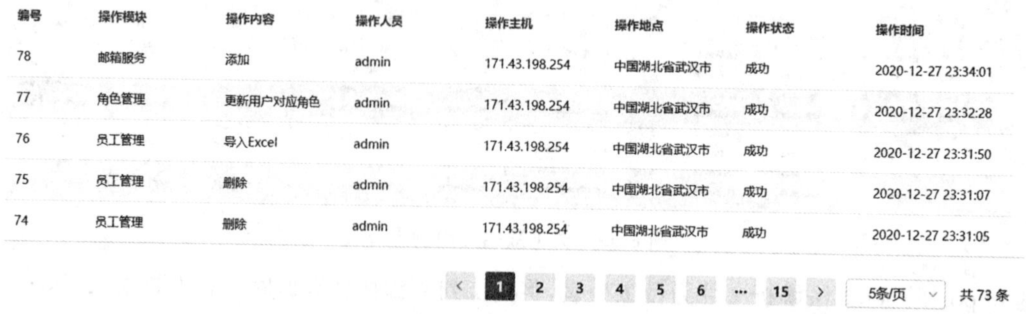

编号	操作模块	操作内容	操作人员	操作主机	操作地点	操作状态	操作时间
78	邮箱服务	添加	admin	171.43.198.254	中国湖北省武汉市	成功	2020-12-27 23:34:01
77	角色管理	更新用户对应角色	admin	171.43.198.254	中国湖北省武汉市	成功	2020-12-27 23:32:28
76	员工管理	导入Excel	admin	171.43.198.254	中国湖北省武汉市	成功	2020-12-27 23:31:50
75	员工管理	删除	admin	171.43.198.254	中国湖北省武汉市	成功	2020-12-27 23:31:07
74	员工管理	删除	admin	171.43.198.254	中国湖北省武汉市	成功	2020-12-27 23:31:05

图 8-37　查看操作日志

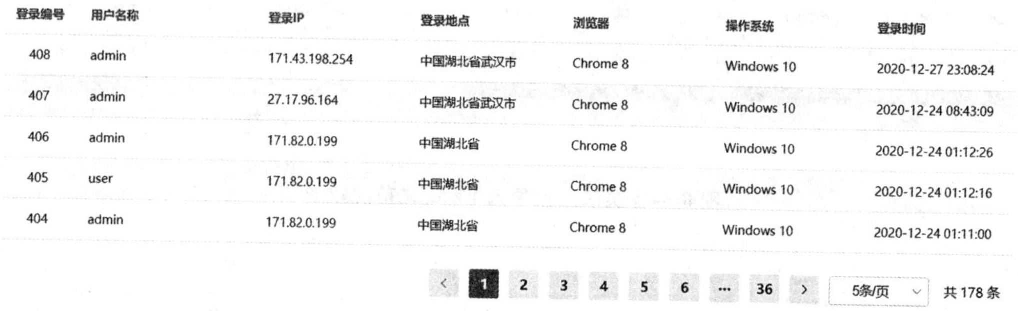

登录编号	用户名称	登录IP	登录地点	浏览器	操作系统	登录时间
408	admin	171.43.198.254	中国湖北省武汉市	Chrome 8	Windows 10	2020-12-27 23:08:24
407	admin	27.17.96.164	中国湖北省武汉市	Chrome 8	Windows 10	2020-12-24 08:43:09
406	admin	171.82.0.199	中国湖北省	Chrome 8	Windows 10	2020-12-24 01:12:26
405	user	171.82.0.199	中国湖北省	Chrome 8	Windows 10	2020-12-24 01:12:16
404	admin	171.82.0.199	中国湖北省	Chrome 8	Windows 10	2020-12-24 01:11:00

图 8-38　查看登录日志

8.4　编码实现

8.4.1　角色管理

1. 数据层 Entity

(1) SysRole.java 文件的源代码如下：

```
package com.easyhao.micro.personnel.entity;
import java.io.Serializable;
public class SysRole implements Serializable {
    private static final long serialVersionUID=5332562598196442940L;
    private Long roleId;
    private String roleName;
    private String roleKey;
    private Boolean delFlag;
    public Long getRoleId() {
        return roleId;
    }
    public void setRoleId(Long roleId) {
        this.roleId=roleId;
```

```java
    }
    public String getRoleName() {
        return roleName;
    }
    public void setRoleName(String roleName) {
        this.roleName=roleName;
    }
    public String getRoleKey() {
        return roleKey;
    }
    public void setRoleKey(String roleKey) {
        this.roleKey=roleKey;
    }
    public Boolean getDelFlag() {
        return delFlag;
    }
    public void setDelFlag(Boolean delFlag) {
        this.delFlag=delFlag;
    }
    @Override
    public String toString() {
        return "SysRole{" +
                "roleId=" + roleId +
                ", roleName='" + roleName + '\'' +
                ", roleKey='" + roleKey + '\'' +
                ", delFlag=" + delFlag +
                '}';
    }
}
```

(2) SysMenu.java 文件的源代码如下：

```java
package com.easyhao.micro.personnel.entity;
import java.io.Serializable;
import java.util.List;
public class SysMenu implements Serializable {
    private static final long serialVersionUID=-8025765487692557295L;
    private Long menuId;
    private String menuName;
    private Long parentId;
    private String url;
    private String path;
    private String component;
    private String icon;
    private Boolean enabled;
```

```java
    //子菜单
    private List<SysMenu> children;
    //菜单对应的角色集合
    private List<SysRole> roles;
    //省略 getter、setter 方法
}
```

2. 数据访问层 Mapper

(1) SysRoleMapper.java 文件的源代码如下:

```java
package com.easyhao.micro.personnel.mapper;
import com.easyhao.micro.personnel.entity.SysRole;
import java.util.List;
public interface SysRoleMapper {
    List<SysRole> selectRolesByUserId(Long id);
    List<SysRole> selectRoleAll();
    int insertRole(SysRole role);
    int deleteRoleById(Long roleId);
    int updateRole(SysRole role);
}
```

(2) SysRoleMapper.xml 文件的源代码如下:

```xml
<?xml version="1.0" encoding="UTF-8"?>
<!DOCTYPE mapper
    PUBLIC "-//mybatis.org//DTD Mapper 3.0//EN"
    "http://mybatis.org/dtd/mybatis-3-mapper.dtd">
<mapper namespace="com.easyhao.micro.personnel.mapper.SysRoleMapper">
    <resultMap id="SysRoleResult" type="com.easyhao.micro.personnel.entity.SysRole">
        <id column="role_id" property="roleId"/>
        <result column="role_name" property="roleName"/>
        <result column="role_key" property="roleKey"/>
        <result column="del_flag" property="delFlag"/>
    </resultMap>
    <sql id="SysRoleColumnAll">
        role_id,role_name,role_key,del_flag
    </sql>
    <select id="selectRolesByUserId" resultMap="SysRoleResult">
        select r.* from sys_role r inner join sys_user_role ur on r.role_id=
            ur.role_id where ur.user_id=#{id}
    </select>
    <select id="selectRoleAll" resultMap="SysRoleResult">
        select <include refid="SysRoleColumnAll"/> from sys_role where del_flag=0
    </select>
    <insert id="insertRole">
        insert into sys_role values (null,#{roleName},#{roleKey},0)
```

```xml
            </insert>
            <update id="deleteRoleById">
                update sys_role set del_flag=1 where role_id=#{roleId}
            </update>
            <update id="updateRole">
                update sys_role set role_name=#{roleName},role_key=
                    #{roleKey} where role_id=#{roleId} and del_flag=0
            </update>
</mapper>
```

（3）SysRoleMenuMapper.java 文件的源代码如下：

```java
package com.easyhao.micro.personnel.mapper;
import org.apache.ibatis.annotations.Param;
public interface SysRoleMenuMapper {
    int deleteRoleMenuByRoleId(Long roleId);
    int insertRoleMenu(@Param("roleId") Long roleId,
        @Param("menuIds") Long[] menuIds);
}
```

（4）SysRoleMenuMapper.xml 文件的源代码如下：

```xml
<?xml version="1.0" encoding="UTF-8"?>
<!DOCTYPE mapper
    PUBLIC "-//mybatis.org//DTD Mapper 3.0//EN"
    "http://mybatis.org/dtd/mybatis-3-mapper.dtd">
<mapper namespace="com.easyhao.micro.personnel.mapper.SysRoleMenuMapper">
    <delete id="deleteRoleMenuByRoleId">
        delete from sys_role_menu where role_id=#{roleId}
    </delete>
    <insert id="insertRoleMenu">
        insert into sys_role_menu (role_id, menu_id) values
        <foreach collection="menuIds" separator="," item="menuId">
            (#{roleId},#{menuId})
        </foreach>
    </insert>
</mapper>
```

3. 业务逻辑层 Service

（1）ISysRoleService.java 文件的源代码如下：

```java
package com.easyhao.micro.personnel.service;
import com.easyhao.micro.personnel.entity.SysRole;
import java.util.List;
public interface ISysRoleService {
    List<SysRole> selectRolesByUserId(Long id);
    List<SysRole> selectRoleAll();
```

```java
    int insertRole(SysRole role);
    int deleteRoleById(Long roleId);
    int updateRole(SysRole role);
}
```

（2）SysRoleServiceImpl.java 文件的源代码如下：

```java
package com.easyhao.micro.personnel.service.impl;
import com.easyhao.micro.personnel.entity.SysRole;
import com.easyhao.micro.personnel.mapper.SysRoleMapper;
import com.easyhao.micro.personnel.mapper.SysRoleMenuMapper;
import com.easyhao.micro.personnel.service.ISysRoleService;
import org.springframework.beans.factory.annotation.Autowired;
import org.springframework.stereotype.Service;
import org.springframework.transaction.annotation.Transactional;
import java.util.List;
@Service
public class SysRoleServiceImpl implements ISysRoleService {
    @Autowired
    SysRoleMapper roleMapper;
    @Autowired
    SysRoleMenuMapper roleMenuMapper;
    public List<SysRole> selectRolesByUserId(Long id){
        return roleMapper.selectRolesByUserId(id);
    }
    public List<SysRole> selectRoleAll() {
        return roleMapper.selectRoleAll();
    }
    public int insertRole(SysRole role) {
        if (!role.getRoleKey().startsWith("ROLE_")) {
            role.setRoleKey("ROLE_"+role.getRoleKey());
        }
        return roleMapper.insertRole(role);
    }
    @Transactional
    public int deleteRoleById(Long roleId) {
        roleMenuMapper.deleteRoleMenuByRoleId(roleId);
        return roleMapper.deleteRoleById(roleId);
    }
    public int updateRole(SysRole role) {
        if (!role.getRoleKey().startsWith("ROLE_")) {
            role.setRoleKey("ROLE_"+role.getRoleKey());
        }
        return roleMapper.updateRole(role);
    }
}
```

}

(3) ISysRoleMenuService.java 文件的源代码如下：

```java
package com.easyhao.micro.personnel.service;
public interface ISysRoleMenuService {
    boolean updateRoleMenu(Long roleId, Long[] menuIds);
}
```

(4) SysRoleMenuServiceImpl.java 文件的源代码如下：

```java
package com.easyhao.micro.personnel.service.impl;
import com.easyhao.micro.personnel.mapper.SysRoleMenuMapper;
import com.easyhao.micro.personnel.service.ISysRoleMenuService;
import org.springframework.beans.factory.annotation.Autowired;
import org.springframework.stereotype.Service;
import org.springframework.transaction.annotation.Transactional;
@Service
public class SysRoleMenuServiceImpl implements ISysRoleMenuService {
    @Autowired
    SysRoleMenuMapper roleMenuMapper;
    @Transactional
    public boolean updateRoleMenu(Long roleId, Long[] menuIds) {
        //roleId 为空，即不需要删除，i=0
        roleMenuMapper.deleteRoleMenuByRoleId(roleId);
        return roleMenuMapper.insertRoleMenu(roleId, menuIds)==menuIds.length;
    }
}
```

4. 控制层 Controller

SysRoleController.java 文件的源代码如下：

```java
package com.easyhao.micro.personnel.controller.sys;
import com.easyhao.micro.personnel.entity.SysRole;
import com.easyhao.micro.personnel.service.ISysRoleMenuService;
import com.easyhao.micro.personnel.service.ISysRoleService;
import com.easyhao.micro.personnel.utils.JsonResult;
import com.easyhao.micro.personnel.aop.Log;
import org.springframework.beans.factory.annotation.Autowired;
import org.springframework.web.bind.annotation.*;
import java.util.List;
@RestController
@RequestMapping("/sys/role")
public class SysRoleController {
    @Autowired
    ISysRoleService roleService;
    @Autowired
```

```java
ISysRoleMenuService roleMenuService;
@GetMapping("/")
public List<SysRole> roleAll() {
    return roleService.selectRoleAll();
}
@PostMapping("/")
@Log(module="角色管理", operContent="添加")
public JsonResult insert(@RequestBody SysRole role) {
    if (roleService.insertRole(role)==1) {
        return JsonResult.success("角色添加成功!");
    }
    return JsonResult.error("角色添加失败!");
}
@DeleteMapping("/{roleId}")
@Log(module="角色管理", operContent="删除")
public JsonResult delete(@PathVariable Long roleId) {
    if (roleService.deleteRoleById(roleId)==1) {
        return JsonResult.success("角色删除成功!");
    }
    return JsonResult.error("角色删除失败!");
}
@PutMapping("/")
@Log(module="角色管理", operContent="更新")
public JsonResult update(@RequestBody SysRole role) {
    if (roleService.updateRole(role)==1) {
        return JsonResult.success("角色更新成功!");
    }
    return JsonResult.error("角色更新失败!");
}
/**
 * 更新 sys_role_menu
 */
@PutMapping("/role_menu")
@Log(module="角色管理", operContent="更新角色对应菜单")
public JsonResult updateRoleMenu(Long roleId, Long[] menuIds) {
    if (roleMenuService.updateRoleMenu(roleId, menuIds)) {
        return JsonResult.success("角色菜单更新成功!");
    }
    return JsonResult.error("角色菜单更新失败!");
}
}
```

8.4.2 部门管理

1. 数据层 Entity

SysDept.java 文件的源代码如下：

```java
package com.easyhao.micro.personnel.entity;
import cn.afterturn.easypoi.excel.annotation.Excel;
import cn.afterturn.easypoi.excel.annotation.ExcelTarget;
import com.fasterxml.jackson.annotation.JsonFormat;
import java.io.Serializable;
import java.sql.Timestamp;
import java.util.List;
@ExcelTarget("SysDept")
public class SysDept implements Serializable {
    private static final long serialVersionUID=-121690740222490895L;
    private Long deptId;
    private Long parentId;
    private String ancestors;
    @Excel(name="所属部门")
    private String deptName;
    private Timestamp createTime;
    private Timestamp updateTime;
    private List<SysDept> children;
    public List<SysDept> getChildren() {
        return children;
    }
    public void setChildren(List<SysDept> children) {
        this.children=children;
    }
    public Long getDeptId() {
        return deptId;
    }
    public void setDeptId(Long deptId) {
        this.deptId=deptId;
    }
    public Long getParentId() {
        return parentId;
    }
    public void setParentId(Long parentId) {
        this.parentId=parentId;
    }
    public String getAncestors() {
        return ancestors;
    }
```

```java
    public void setAncestors(String ancestors) {
        this.ancestors=ancestors;
    }
    public String getDeptName() {
        return deptName;
    }
    public void setDeptName(String deptName) {
        this.deptName=deptName;
    }
    @JsonFormat(pattern="yyyy-MM-dd HH:mm:ss", timezone="Asia/Shanghai")
    public Timestamp getCreateTime() {
        return createTime;
    }
    public void setCreateTime(Timestamp createTime) {
        this.createTime=createTime;
    }
    @JsonFormat(pattern="yyyy-MM-dd HH:mm:ss", timezone="Asia/Shanghai")
    public Timestamp getUpdateTime() {
        return updateTime;
    }
    public void setUpdateTime(Timestamp updateTime) {
        this.updateTime=updateTime;
    }
    @Override
    public String toString() {
        return "SysDept{" +
            "deptId=" +deptId +
            ", parentId=" +parentId +
            ", ancestors='" +ancestors + '\'' +
            ", deptName='" +deptName + '\'' +
            ", createTime=" +createTime +
            ", updateTime=" +updateTime +
            ", children=" +children +
            '}';
    }
}
```

2. 数据访问层 Mapper

（1）SysDeptMapper.java 文件的源代码如下：

```java
package com.easyhao.micro.personnel.mapper;
import com.easyhao.micro.personnel.entity.SysDept;
import java.util.List;
public interface SysDeptMapper {
    List<SysDept> selectDeptAll(Long deptId);
```

```
    int insertDept(SysDept dept);
    int deleteDeptById(Long deptId);
    int updateDept(SysDept dept);
    //部门是否存在员工
    int hasEmpByDeptId(Long deptId);
    List<SysDept> selectDeptIdAndDeptName();
    Long selectDeptIdByDeptName(String deptName);
}
```

(2) SysDeptMapper.xml 文件的源代码如下：

```xml
<?xml version="1.0" encoding="UTF-8"?>
<!DOCTYPE mapper
    PUBLIC "-//mybatis.org//DTD Mapper 3.0//EN"
    "http://mybatis.org/dtd/mybatis-3-mapper.dtd">
<mapper namespace="com.easyhao.micro.personnel.mapper.SysDeptMapper">
    <resultMap id="SysDeptResult" type="SysDept">
        <id column="dept_id" property="deptId"/>
        <result column="parent_id" property="parentId"/>
        <result column="ancestors" property="ancestors"/>
        <result column="dept_name" property="deptName"/>
        <result column="create_time" property="createTime"/>
        <result column="update_time" property="updateTime"/>
    </resultMap>
    <resultMap id="selectDeptAll" type="SysDept" extends="SysDeptResult">
        <collection property="children" ofType="SysDept" select=
            "selectDeptAll" column="dept_id"/>
    </resultMap>
    <select id="selectDeptAll" resultMap="selectDeptAll">
        select * from sys_dept where parent_id=#{deptId}
    </select>
    <insert id="insertDept" parameterType="SysDept" useGeneratedKeys=
        "true" keyProperty="deptId">
        insert into sys_dept values (null,#{parentId},#{ancestors},
        #{deptName},now(),now());
    </insert>
    <delete id="deleteDeptById">
        delete from sys_dept where dept_id=#{deptId}
    </delete>
    <update id="updateDept">
        update sys_dept set dept_name=#{deptName},update_time=
            now() where dept_id=#{deptId};
    </update>
    <select id="hasEmpByDeptId" resultType="int">
        select count(dept.dept_id) from sys_dept dept inner join sys_emp emp on
```

```xml
            emp.dept_id=dept.dept_id and dept.dept_id=#{deptId};
    </select>
    <resultMap id="selectDeptIdAndDeptName" type="SysDept">
        <result column="dept_id" property="deptId"/>
        <result column="dept_name" property="deptName"/>
    </resultMap>
    <select id="selectDeptIdAndDeptName" resultMap="selectDeptIdAndDeptName">
        select dept.dept_id,dept.dept_name from sys_dept dept;
    </select>
    <select id="selectDeptIdByDeptName" resultType="java.lang.Long">
        select dept_id from sys_dept where dept_name=#{deptName};
    </select>
</mapper>
```

3. 业务逻辑层 Service

（1）SysDeptService.java 文件的源代码如下：

```java
package com.easyhao.micro.personnel.service;
import com.easyhao.micro.personnel.entity.SysDept;
import java.util.List;
public interface ISysDeptService {
    List<SysDept> selectDeptAll();
    int insertDept(SysDept dept);
    int deleteDeptById(Long deptId);
    int updateDept(SysDept dept);
    int hasEmpByDeptId(Long deptId);
    List<SysDept> selectDeptIdAndDeptName();
    Long selectDeptIdByDeptName(String deptName);
}
```

（2）SysDeptServiceImpl.java 文件的源代码如下：

```java
package com.easyhao.micro.personnel.service.impl;
import com.easyhao.micro.personnel.entity.SysDept;
import com.easyhao.micro.personnel.mapper.SysDeptMapper;
import com.easyhao.micro.personnel.service.ISysDeptService;
import org.springframework.beans.factory.annotation.Autowired;
import org.springframework.stereotype.Service;
import org.springframework.transaction.annotation.Transactional;
import java.util.List;
@Service
public class SysDeptServiceImpl implements ISysDeptService {
    @Autowired
    SysDeptMapper deptMapper;
    @Override
    public List<SysDept> selectDeptAll() {
```

```java
        return deptMapper.selectDeptAll(0L);
    }
    @Override
    @Transactional
    public int insertDept(SysDept dept) {
        dept.setParentId(dept.getDeptId());
        dept.setAncestors(dept.getAncestors()+","+dept.getDeptId());
        return deptMapper.insertDept(dept);
    }
    @Override
    public int deleteDeptById(Long deptId) {
        return deptMapper.deleteDeptById(deptId);
    }
    @Override
    public int updateDept(SysDept dept) {
        return deptMapper.updateDept(dept);
    }
    @Override
    public int hasEmpByDeptId(Long deptId) {
        return deptMapper.hasEmpByDeptId(deptId);
    }
    @Override
    public List<SysDept> selectDeptIdAndDeptName() {
        return deptMapper.selectDeptIdAndDeptName();
    }
    @Override
    public Long selectDeptIdByDeptName(String deptName) {
        return deptMapper.selectDeptIdByDeptName(deptName);
    }
}
```

4. 控制层 Controller

SysDeptController.java 文件的源代码如下：

```java
package com.easyhao.micro.personnel.controller.sys;
import com.easyhao.micro.personnel.entity.SysDept;
import com.easyhao.micro.personnel.service.ISysDeptService;
import com.easyhao.micro.personnel.utils.JsonResult;
import com.easyhao.micro.personnel.aop.Log;
import org.springframework.beans.factory.annotation.Autowired;
import org.springframework.web.bind.annotation.*;
import java.util.List;
@RestController
@RequestMapping("/sys/dept")
public class SysDeptController {
```

```java
@Autowired
ISysDeptService deptService;
@GetMapping("/")
public List<SysDept> list() {
    return deptService.selectDeptAll();
}
@GetMapping("/idAndName")
public List<SysDept> deptIdAndName() {
    return deptService.selectDeptIdAndDeptName();
}
@PostMapping("/")
@Log(module="部门管理", operContent="添加")
public JsonResult insert(@RequestBody SysDept dept) {
    if (deptService.insertDept(dept)==1) {
        return JsonResult.success("添加部门成功!");
    }
    return JsonResult.error("添加部门失败!");
}
@DeleteMapping("/{deptId}")
@Log(module="部门管理", operContent="删除")
public JsonResult delete(@PathVariable Long deptId) {
    //先判断该部门下是否存在员工
    if (deptService.hasEmpByDeptId(deptId)>0) {
        return JsonResult.error("该部门存在员工,无法删除!");
    }
    if (deptService.deleteDeptById(deptId)==1) {
        return JsonResult.success("部门删除成功!");
    }
    return JsonResult.error("部门删除失败!");
}
@PutMapping("/")
@Log(module="部门管理", operContent="更新")
public JsonResult update(@RequestBody SysDept dept) {
    if (deptService.updateDept(dept)==1) {
        return JsonResult.success("更新部门成功!");
    }
    return JsonResult.error("更新部门失败!");
}
}
```

8.4.3 员工管理

1. 数据层 Entity

SysEmp.java 文件的源代码如下：

```java
package com.easyhao.micro.personnel.entity;
import cn.afterturn.easypoi.excel.annotation.Excel;
import cn.afterturn.easypoi.excel.annotation.ExcelEntity;
import cn.afterturn.easypoi.excel.annotation.ExcelIgnore;
import cn.afterturn.easypoi.excel.annotation.ExcelTarget;
import com.fasterxml.jackson.annotation.JsonFormat;
import java.io.Serializable;
import java.sql.Date;
import java.sql.Timestamp;
@ExcelTarget("SysEmp")
public class SysEmp implements Serializable {
    private static final long serialVersionUID=1686636778395502136L;
    @Excel(name="员工编号")
    private Long empId;
    @ExcelIgnore
    private Long deptId;
    @Excel(name="员工姓名")
    private String empName;
    @ExcelEntity
    private SysDept dept;
    @Excel(name="性别")
    private String gender;
    @Excel(name="出生日期", format="yyyy-MM-dd", width=20)
    private Date birthday;
    @Excel(name="籍贯", width=20)
    private String nativePlace;
    @Excel(name="民族")
    private String nation;
    @Excel(name="身份证号", width=30)
    private String idCard;
    @Excel(name="手机号", width=20)
    private String phone;
    @Excel(name="邮箱", width=25)
    private String email;
    @Excel(name="学位")
    private String degree;
    @Excel(name="职位", width=20)
    private String position;
    @Excel(name="任职时间", format="yyyy-MM-dd", width=20)
    private Date workTime;
    @Excel(name="创建时间", format="yyyy-MM-dd HH:mm:ss", width=25)
    private Timestamp createTime;
    @Excel(name="更新时间", format="yyyy-MM-dd HH:mm:ss", width=25)
    private Timestamp updateTime;
```

```java
    public SysDept getDept() {
        return dept;
    }
    public void setDept(SysDept dept) {
        this.dept=dept;
    }
    public Long getEmpId() {
        return empId;
    }
    public void setEmpId(Long empId) {
        this.empId=empId;
    }
    public Long getDeptId() {
        return deptId;
    }
    public void setDeptId(Long deptId) {
        this.deptId=deptId;
    }
    public String getEmpName() {
        return empName;
    }
    public void setEmpName(String empName) {
        this.empName=empName;
    }
    public String getGender() {
        return gender;
    }
    public void setGender(String gender) {
        this.gender=gender;
    }
    @JsonFormat(pattern="yyyy-MM-dd", timezone="Asia/Shanghai")
    public Date getBirthday() {
        return birthday;
    }
    public void setBirthday(Date birthday) {
        this.birthday=birthday;
    }
    public String getNativePlace() {
        return nativePlace;
    }
    public void setNativePlace(String nativePlace) {
        this.nativePlace=nativePlace;
    }
    public String getNation() {
```

```java
        return nation;
    }
    public void setNation(String nation) {
        this.nation=nation;
    }
    public String getIdCard() {
        return idCard;
    }
    public void setIdCard(String idCard) {
        this.idCard=idCard;
    }
    public String getPhone() {
        return phone;
    }
    public void setPhone(String phone) {
        this.phone=phone;
    }
    public String getEmail() {
        return email;
    }
    public void setEmail(String email) {
        this.email=email;
    }
    public String getDegree() {
        return degree;
    }
    public void setDegree(String degree) {
        this.degree=degree;
    }
    public String getPosition() {
        return position;
    }
    public void setPosition(String position) {
        this.position=position;
    }
    @JsonFormat(pattern="yyyy-MM-dd", timezone="Asia/Shanghai")
    public Date getWorkTime() {
        return workTime;
    }
    public void setWorkTime(Date workTime) {
        this.workTime=workTime;
    }
    @JsonFormat(pattern="yyyy-MM-dd HH:mm:ss", timezone="Asia/Shanghai")
    public Timestamp getCreateTime() {
```

```java
            return createTime;
        }
        public void setCreateTime(Timestamp createTime) {
            this.createTime=createTime;
        }
        @JsonFormat(pattern="yyyy-MM-dd HH:mm:ss", timezone="Asia/Shanghai")
        public Timestamp getUpdateTime() {
            return updateTime;
        }
        public void setUpdateTime(Timestamp updateTime) {
            this.updateTime=updateTime;
        }
        @Override
        public String toString() {
            return "SysEmp{" +
                    "empId=" + empId +
                    ", deptId=" + deptId +
                    ", empName='" + empName + '\'' +
                    ", dept=" + dept +
                    ", gender='" + gender + '\'' +
                    ", birthday=" + birthday +
                    ", nativePlace='" + nativePlace + '\'' +
                    ", nation='" + nation + '\'' +
                    ", idCard='" + idCard + '\'' +
                    ", phone='" + phone + '\'' +
                    ", email='" + email + '\'' +
                    ", degree='" + degree + '\'' +
                    ", position='" + position + '\'' +
                    ", workTime=" + workTime +
                    ", createTime=" + createTime +
                    ", updateTime=" + updateTime +
                    '}';
        }
    }
}
```

2. 数据访问层 Mapper

（1）SysEmpMapper.java 文件的源代码如下：

```java
package com.easyhao.micro.personnel.mapper;
import com.easyhao.micro.personnel.entity.SysEmp;
import org.apache.ibatis.annotations.Param;
import java.util.List;
public interface SysEmpMapper {
    List<SysEmp> selectEmpList(SysEmp emp);
    int deleteEmpById(Long empId);
```

```
    int insertEmp(SysEmp emp);
    int updateEmp(SysEmp emp);
    int insertEmpList(@Param("emps") List<SysEmp> emps);
}
```

(2) SysEmpMapper.xml 文件的源代码如下:

```xml
<?xml version="1.0" encoding="UTF-8"?>
<!DOCTYPE mapper
        PUBLIC "-//mybatis.org//DTD Mapper 3.0//EN"
        "http://mybatis.org/dtd/mybatis-3-mapper.dtd">
<mapper namespace="com.easyhao.micro.personnel.mapper.SysEmpMapper">
    <resultMap id="SysEmpResultMap" type=
        "com.easyhao.micro.personnel.entity.SysEmp">
        <id column="emp_id" property="empId"/>
        <result column="dept_id" property="deptId"/>
        <result column="emp_name" property="empName"/>
        <result column="gender" property="gender"/>
        <result column="birthday" property="birthday"/>
        <result column="native_place" property="nativePlace"/>
        <result column="nation" property="nation"/>
        <result column="id_card" property="idCard"/>
        <result column="phone" property="phone"/>
        <result column="email" property="email"/>
        <result column="degree" property="degree"/>
        <result column="position" property="position"/>
        <result column="work_time" property="workTime"/>
        <result column="create_time" property="createTime"/>
        <result column="update_time" property="updateTime"/>
    </resultMap>
    <resultMap id="selectEmpListMap" type="SysEmp" extends="SysEmpResultMap">
        <association property="dept" javaType="com.easyhao.micro.personnel.entity.SysDept">
            <result column="dept_name" property="deptName"/>
        </association>
    </resultMap>
    <select id="selectEmpList" resultMap="selectEmpListMap" parameterType="SysEmp">
        select emp.*,dept.dept_name from sys_emp emp
            inner join sys_dept dept on emp.dept_id=dept.dept_id
        <where>
            <if test="empName !=null and empName !=''">
                emp.emp_name like concat('%',#{empName},'%')
            </if>
            <if test="dept !=null and dept.deptName !=null and dept.deptName !=''">
                and dept.dept_name like concat('%',#{dept.deptName},'%')
            </if>
            <if test="gender !=null and gender !=''">
                and emp.gender=#{gender}
            </if>
```

```xml
            </where>
        </select>
        <delete id="deleteEmpById">
            delete from sys_emp where emp_id=#{empId}
        </delete>
        <insert id="insertEmp" parameterType="SysEmp">
            insert into sys_emp values
            (
                null,#{deptId},#{empName},#{gender},
                #{birthday},#{nativePlace},#{nation},#{idCard},
                #{phone},#{email},#{degree},#{position},#{workTime},now(),now()
            );
        </insert>
        <update id="updateEmp" parameterType="SysEmp">
            update sys_emp set
            dept_id=#{deptId},emp_name=#{empName},gender=#{gender},
                birthday=#{birthday},native_place=#{nativePlace},nation=
                #{nation},id_card=#{idCard},
            phone=#{phone},email=#{email},degree=#{degree},position=#{position},
                work_time=#{workTime},update_time=now()
                where emp_id=#{empId};
        </update>
        <insert id="insertEmpList">
            insert into sys_emp values
            <foreach collection="emps" item="emp" separator=",">
                (
                null,
                #{emp.deptId},#{emp.empName},#{emp.gender},
                #{emp.birthday},#{emp.nativePlace},#{emp.nation},
                #{emp.idCard},#{emp.phone},#{emp.email},
                #{emp.degree},#{emp.position},#{emp.workTime},
                #{emp.createTime},#{emp.updateTime}
                )
            </foreach>
        </insert>
</mapper>
```

3. 业务逻辑层 Service

(1) ISysEmpService.java 文件的源代码如下：

```
package com.easyhao.micro.personnel.service;
import com.easyhao.micro.personnel.entity.SysEmp;
import java.util.List;
public interface ISysEmpService {
    List<SysEmp> selectEmpList(SysEmp emp);
    int deleteEmpById(Long empId);
    int insertEmp(SysEmp emp);
    int updateEmp(SysEmp emp);
```

```
    int insertEmpListByExcel(List<SysEmp> emps);
}
```

（2）SysEmpServiceImpl 文件的源代码如下：

```java
package com.easyhao.micro.personnel.service.impl;
import com.easyhao.micro.personnel.entity.SysEmp;
import com.easyhao.micro.personnel.mapper.SysEmpMapper;
import com.easyhao.micro.personnel.service.ISysDeptService;
import com.easyhao.micro.personnel.service.ISysEmpService;
import org.springframework.beans.factory.annotation.Autowired;
import org.springframework.stereotype.Service;
import java.util.List;
@Service
public class SysEmpServiceImpl implements ISysEmpService {
    @Autowired
    SysEmpMapper empMapper;
    @Autowired
    ISysDeptService deptService;
    @Override
    public List<SysEmp> selectEmpList(SysEmp emp) {
        return empMapper.selectEmpList(emp);
    }
    @Override
    public int deleteEmpById(Long empId) {
        return empMapper.deleteEmpById(empId);
    }
    @Override
    public int insertEmp(SysEmp emp) {
        return empMapper.insertEmp(emp);
    }
    @Override
    public int updateEmp(SysEmp emp) {
        return empMapper.updateEmp(emp);
    }
    @Override
    public int insertEmpListByExcel(List<SysEmp> emps) {
        for (SysEmp emp : emps) {
            Long deptId=deptService.selectDeptIdByDeptName(
                emp.getDept().getDeptName());
            emp.setDeptId(deptId);
        }
        //集合添加到数据库
        return empMapper.insertEmpList(emps);
    }
}
```

}

4. 控制层 Controller

SysEmpController 文件的源代码如下：

```java
package com.easyhao.micro.personnel.controller.sys;
import cn.afterturn.easypoi.excel.ExcelExportUtil;
import cn.afterturn.easypoi.excel.ExcelImportUtil;
import cn.afterturn.easypoi.excel.entity.ExportParams;
import cn.afterturn.easypoi.excel.entity.ImportParams;
import com.easyhao.micro.personnel.entity.SysEmp;
import com.easyhao.micro.personnel.service.ISysEmpService;
import com.easyhao.micro.personnel.utils.JsonResult;
import com.easyhao.micro.personnel.aop.Log;
import org.apache.poi.ss.usermodel.Workbook;
import org.springframework.beans.factory.annotation.Autowired;
import org.springframework.http.HttpHeaders;
import org.springframework.http.HttpStatus;
import org.springframework.http.MediaType;
import org.springframework.http.ResponseEntity;
import org.springframework.web.bind.annotation.*;
import org.springframework.web.multipart.MultipartFile;
import java.io.ByteArrayOutputStream;
import java.io.IOException;
import java.nio.charset.StandardCharsets;
import java.util.List;
@RestController
@RequestMapping("/sys/emp")
public class SysEmpController {
    @Autowired
    ISysEmpService empService;
    @PostMapping("/list")
    public List<SysEmp> list(@RequestBody SysEmp emp) {
        return empService.selectEmpList(emp);
    }
    @DeleteMapping("/{empId}")
    @Log(module="员工管理", operContent="删除")
    public JsonResult delete(@PathVariable Long empId) {
        if (empService.deleteEmpById(empId)==1) {
            return JsonResult.success("员工删除成功!");
        }
        return JsonResult.error("员工删除失败!");
    }
    @PostMapping("/")
    @Log(module="员工管理", operContent="添加")
```

```java
public JsonResult insert(@RequestBody SysEmp emp) {
    if (empService.insertEmp(emp)==1) {
        return JsonResult.success("员工添加成功!");
    }
    return JsonResult.error("员工添加失败!");
}
@PutMapping("/")
@Log(module="员工管理", operContent="更新")
public JsonResult update(@RequestBody SysEmp emp) {
    if (empService.updateEmp(emp)==1) {
        return JsonResult.success("员工更新成功!");
    }
    return JsonResult.error("员工更新失败!");
}
/**
 *下载文件,在 Spring MVC 中,返回 ResponseEntity
 */
@GetMapping("/export")
@Log(module="员工管理", operContent="导出 Excel")
public ResponseEntity<byte[]> exportExcel() {
    List<SysEmp> empList=empService.selectEmpList(null);
    Workbook workbook=ExcelExportUtil.exportExcel(
        new ExportParams(), SysEmp.class, empList);
    ByteArrayOutputStream baos=new ByteArrayOutputStream();
    HttpHeaders headers=new HttpHeaders();
    try {
        headers.setContentDispositionFormData("attachment",
            new String("员工列表.xls".getBytes(StandardCharsets.UTF_8),
            StandardCharsets.ISO_8859_1));
        headers.setContentType(MediaType.APPLICATION_OCTET_STREAM);
        workbook.write(baos);
    } catch (IOException e) {
        e.printStackTrace();
    }
    return new ResponseEntity<>(baos.toByteArray(), headers,
        HttpStatus.CREATED);
}
@PostMapping("/import")
@Log(module="员工管理", operContent="导入 Excel")
public JsonResult importExcel(MultipartFile file) throws Exception {
    ImportParams importParams=new ImportParams();
    importParams.setTitleRows(0);
    importParams.setHeadRows(1);
    List<SysEmp> list=ExcelImportUtil.importExcel(
```

```
            file.getInputStream(), SysEmp.class, importParams);
        int i=empService.insertEmpListByExcel(list);
        if (i>0) {
            return JsonResult.success("Excel 导入成功!");
        } else {
            return JsonResult.error("Excel 导入失败!");
        }
    }
}
```

8.5 项目搭建

员工管理系统使用 IntelliJ IDEA 工具和 MySQL 数据库开发,系统项目结构如图8-39、图 8-40 所示。

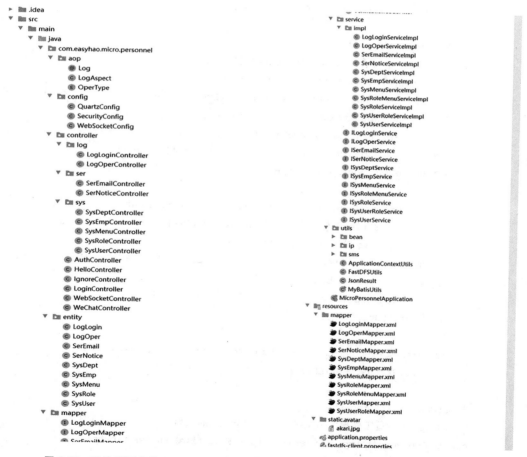

图 8-39　系统项目结构图 1　　　　　　图 8-40　系统项目结构图 2

参 考 文 献

[1] 罗旋.Java Web应用开发教程[M].武汉:华中科技大学出版社,2020.
[2] 明日科技.Java Web项目开发实战入门[M].吉林:吉林大学出版社,2017.
[3] 陈学明.Spring+Spring MVC+MyBatis整合开发实战[M].北京:机械工业出版社,2020.
[4] 肖海鹏,牟东旭.SSM与Spring Boot开发实战[M].北京:人民邮电出版社,2020.
[5] 缪勇,施俊.Spring+Spring MVC+MyBatis框架技术精讲与整合案例[M].北京:清华大学出版社,2019.